A Practical Handbook on Measurement Uncertainty

Faqs and fundamentals for metrologists

Online at: https://doi.org/10.1088/978-0-7503-6462-1

A Practical Handbook on Measurement Uncertainty

FAQs and fundamentals for metrologists

Swanand Rishi

IOP Publishing, Bristol, UK

ISBN   978-0-7503-6462-1 (ebook)
ISBN   978-0-7503-6458-4 (print)
ISBN   978-0-7503-6459-1 (myPrint)
ISBN   978-0-7503-6461-4 (mobi)

DOI   10.1088/978-0-7503-6462-1

Version: 20240701

IOP ebooks

British Library Cataloguing-in-Publication Data: A catalogue record for this book is available from the British Library.

Published by IOP Publishing, wholly owned by The Institute of Physics, London

IOP Publishing, No.2 The Distillery, Glassfields, Avon Street, Bristol, BS2 0GR, UK

US Office: IOP Publishing, Inc., 190 North Independence Mall West, Suite 601, Philadelphia, PA 19106, USA

Dedicated to my parents,

the late Nagesh Rishi and Sukhada Rishi

for their noble parenting;

and

my wife

Madhura Rishi

for her support and encouragement.

Declaration

The views expressed in this book are those of the author and not those of the
Government of India.

Contents

Preface

Doubt is not a pleasant condition, but certainty is absurd.

—Voltaire

Uncertainty is the order of the day in practically every sphere of activity. It is commonplace in political, social, cultural, financial, and economic domains worldwide. The globalization of trade, fierce competition, outsourcing of products and services, technological advances, and a mix of all the above contribute to the uncertainty in our social lives. VUCA (Volatile, Uncertain, Complex, and Ambiguous) is the buzzword used in management parlance to describe today's competitive business environment. At the personal level, life is also becoming more and more uncertain. One is not sure whether one's smartphone bought today will be there in the market after six months! There are some well-known adages that proclaim 'change is the only constant,' 'the more you change, the more you remain constant,' or 'death and taxes are the only certainties'! To above list, we can add 'uncertainty is the only certainty!' and it is promulgating at an ever-increasing speed. Due to fierce competition in all manufacturing industries, material cost-cutting has gained significance and correct measurements play a vital role there. The stringent regulations, awareness of customers' rights, and the proactive judicial and legislative approach to the redressal of consumer grievances has put a lot of pressure on industries. For instance, the worldwide recall of passenger vehicles by some very renowned manufacturers has cost them dearly in terms of financial losses and goodwill. High accuracy, low uncertainty, guard banding, and appropriate false-accept (or false-reject) criteria, among others, would probably be the solutions to sustain in the aggressive competition in all fields of industry. The service industry, including laboratory services, is no exception. Due to computerization, networking, and the Internet, vast amounts of data (in the form of information) are at one's disposal.

In the field of metrology, uncertainty has a special place of its own. For all laboratory personnel—whether working in testing or calibration (and now sampling laboratories also, as per ISO/IEC 17025:2017)—measurement uncertainty evaluation is part and parcel of their workload. *No measurement result is complete without an explicit statement of its uncertainty.* The measurement uncertainty is estimated as per the Guide to the Expression of Uncertainty in Measurement (GUM) (ISO/IEC Guide 98-3)—the apex document in this field. Its methodology is well established and followed the world over by consensus.

Laboratory personnel receive training in measurement uncertainty evaluation, but the conventional training has two shortcomings: first, it is restricted to imparting working knowledge only and the rationale behind the concepts is generally not explained. The training programmes extend to a day or two, and touching upon all the finer aspects is rather difficult. In addition, participants find it difficult to digest statistical jargon. This may partly be due to the fact that the participants—most of who are at working level or junior level—are not receptive enough to understand the statistical clichés and are interested in quick-fix solutions. Nevertheless, for senior-level metrologists, it is very important to understand these concepts so that measurement uncertainty evaluation is not just a safe, simple, and mundane activity, but a meaningful one. It is important for them to understand that there are no straitjacket solutions in measurement uncertainty evaluation. And this becomes more imperative when conformity decisions are involved.

Second, the training does not necessarily make the participants well versed in the subject, since measurement uncertainty evaluation is *per se* not a straightforward process. The uncertainty components for the same measurand in two scenarios could be different and would lead to altogether different answers. As I usually quote in seminars on this subject, in mathematics, 1 *plus* 1=2; but in statistics, it may or may or may not be so; because in latter, we are confronted with limited data and many assumptions. In measurement, we deal with what is called the 'Strong Law of Small numbers,' (formulated by the British mathematician Richard Guy) which states that *'There aren't enough small numbers to meet the many demands made of them!'* If that is the situation with arithmetic, will it not be more cumbersome in statistics, which is based on a small sample size and a great deal of 'degree of belief'?

When participants finish their training, in spite of the stress placed on the above facts as well as on the importance of critical thinking, they tend to follow the lines of the examples and case studies illustrated during the seminar, forgetting the essence of uncertainty evaluation. They estimate uncertainty mechanically using either spreadsheets or ready-made software without bothering about the fundamentals.

Here, it is worth invoking clause 3.4.8 of the GUM, which states:

> Although this Guide provides a framework for assessing uncertainty, it cannot substitute for ***critical thinking*** (italics mine), intellectual honesty and professional skill. The evaluation of uncertainty is neither a routine task nor a purely mathematical one; it depends on detailed knowledge of the nature of the measurand and of the measurement. The quality and utility of the uncertainty quoted for the result of a measurement therefore ultimately depend on the understanding, critical analysis, and integrity of those who contribute to the assignment of its value.

It is the *critical thinking* that this book is intended to promote and instil in metrologists.

One more observation during laboratory accreditation audits is that it is becoming more and more difficult for auditors to find nonconformities (although it is stated in opening meetings that they are not there to do so!), particularly in laboratories which have been well established and accredited for years. The auditors then tend to probe the knowledge and basic concepts of the auditee, and if found inadequate, recommend a training to clarify the concepts! This book is meant to help the reader understand the concepts.

I am aware that the GUM, its supplements, some national standards, and a few books have elaborated on these aspects. Most of the books, although few in numbers, are quite involved and exhaustive. They are not meant to be handy nor quick reference materials. The fundamental concepts in them are sparse and scattered. This book intends to address these issues at a glance. So, it will be a quick reference book on uncertainty, with subtitles such as 'What is the significance of the sensitivity coefficient?' or 'What are degrees of freedom?' The interrogative titles will invoke the interest and curiosity of the reader.

Apart from books, many standards and guides published by national and international bodies are also available. However, many of these have presented the concepts and the philosophy from an academic perspective. Metrologists (at all levels—by and large), for whatever reasons, generally keep away from these standards and guides. Practitioners are not interested in the derivations of formulae and the intricacies of calculus.

In view of the above, I felt the need of a book intended to:
  (a) clarify the rationale behind certain assumptions or ideas,
  (b) present the concepts from a practical perspective,
  (c) help the reader arrive at a realistic uncertainty figure, and
  (d) be a ready-reference handbook.

In essence, the purpose of the book is to explain the fundamental concepts of measurement uncertainty and the preliminary statistics involved in its implementation. Many of the topics are the result of inquisitive participants in the seminars/ training arranged by my parent organization, ETDC (Pune) under the Standardisation Testing and Quality Certification (STQC) Directorate, Ministry of Electronics and Information Technology, Government of India. This book is general in nature and is not meant for any specific area, audience, or field of metrology. However, I have included some advanced, but elementary topics for those who want to explore beyond the basic GUM philosophy.

Statistics is usually not a major (and also not sought after!) subject in science education, and graduates in science or engineering find it difficult to digest statistical jargon. The terminology is difficult to fathom, and except for those working in laboratories and quality control, working professionals hardly apply it in routine work. For metrologists, it is now obligatory to familiarize themselves with the basics of statistics. Keeping this in mind, however, only essential statistics has been included in this book.

In engineering syllabi also, topics in measurement and calibration are now well established. Nonetheless, these subjects get less attention, as they are regarded as supplementary in nature. I have been a visiting guest lecturer in engineering colleges where the academic professionals have frequently voiced the need for a book that focuses on the essence of measurements and related concepts. Thus, academic staff will also gain from this book.

Apart from metrologists, who are the main audience, I hope the book will also help the laboratory-auditing fraternity, academics, and quality professionals at large in all areas.

Foreword

In their daily professional life, metrologists often look for quick answers to queries that are related to science and technology. This is particularly evident for metrologists who have to deal with issues related to the evaluation of measurement uncertainty. While online resources do help, many times the authenticity of the non-peer reviewed resources without proper references, is questionable. On the other end, we have well established reference books which are expositions and require time and effort to go through. For example, ISO Guide 98-3 (GUM) and its supplements, national standards and a few books are quite involved and exhaustive. And they are not meant to be handy nor quick reference materials.

This carefully written short book by Swanand Rishi fills the gap, both as a refresher course and as a quick reference guide, that tries to explain many fundamental concepts related to the evaluation of measurement uncertainty. This book, titled *A Practical Handbook on Measurement Uncertainty: FAQs and fundamentals for metrologists*, true to its title, handles the measurement uncertainty right from the basics. This would be very useful for beginners as well as seasoned metrologists. It covers various concepts that are often used by metrologists, assessors of accredited laboratories, and technologists in laboratory management and legal metrology. The vast experience of the author is apparent in the clear and succinct manner in which the topics are covered. The content under the 'Facts one should know' and 'Points to ponder' provide more insight and guidance.

This book is very well structured in six different parts. The first four cover the fundamentals, distributions, sample size and analysis, and how to understand the degrees of freedom. The next two parts cover the related topics and a deeper discussion of the topics including how to detect outliers in sample analysis, reporting of uncertainty, and alternative approaches in the evaluation of uncertainty. The inquisitive titles of the chapters certainly invoke interest and curiosity of the reader. The author has very effectively explained the **rationale behind various assumptions or ideas and presented the concepts from a utilitarian perspective**.

As a practitioner with long experience, the author is well versed with the nuances of the field and has handled the subject in a way that is of practical significance. This will be a handy book for practicing and aspiring metrologists, assessors, and laboratory technicians.

Professor Venu Gopal Achanta
Director of National Physical Laboratory (NPL), New Delhi, India
May 2024

Acknowledgements

At the very outset, I am grateful to Prof. Venu Gopal Achanta, who is Director of the National Physical Laboratory (NPL), India. He holds a PhD in Physics for his work on exciton dynamics in low-dimensional semiconductors and a PhD in electronic engineering from Tokyo University for his work on the design and demonstration of an ultrafast all-optical switch. He was a JST Fellow in the Quantum Information Technology group, Basic Research Labs, NEC, Japan. His research interest is classical and quantum information processing. I am honored to have a foreword written by an eminent and distinguished scientist like him.

I would like to acknowledge the contribution of some colleagues and numerous participants in the training courses on 'Measurement Uncertainty,' 'Certified Calibration Professional,' and 'Statistical Process Control' conducted by STQC, for which I have been a trainer for more than two decades. Many of them have been core metrologists in multitude of disciplines, which made me think about issues in the evaluation of measurement uncertainty from different perspectives. It was a wonderful experience to address their queries about practical difficulties and the fundamentals of statistics and measurement uncertainty. My interactions with metrlogists during conferences on metrology revealed the necessity for a book to address fundamental queries on this subject. This certainly prompted me write this book, as if it were my bona fide duty.

I am also thankful to the reviewers of IOP Publishing Ltd for their critical reviews, suggestions, and recommendations which indeed improved the presentation of this book. My special thanks also go to John Navas, Senior Commissioning Manager and his team at IOP Publishing for their wonderful and patient guidance on taking this book to publication.

Last but not least, my heartfelt thanks are due to my wife Madhura—a working professional herself—and daughters Sanika and Prachi for being supportive as well as forbearing during the period of preparation of this book.

Swanand Rishi

Author biography

Swanand Rishi

Academic and professional:

Swanand Rishi graduated in electrical engineering with distinction in 1986 (with specialization in power electronics) from College of Engineering, Pune, Maharashtra, India.

Diploma in Business Management (DBM) in 1998, with Gold medal.

ISO 9001 Assessor.

Assessor for IT Test Laboratories.

Work experience:

After graduation in 1986, he worked in a few industries that manufactured uninterruptible power supplies (UPSs), power conditioners, capacitors, and lightning arrestors. He worked as project manager in a company that supplied UPSs for a missile system of the Indian army. He was involved in R&D at the process level and was instrumental in drastically reducing the rejections of certain types of capacitors. He had developed many electronic gadgets for testing of lightning arrestors that helped assure the quality of tested products.

Since 1995, he has been working in Electronics Test & Development Centre (ETDC), a test and calibration laboratory of the STQC Directorate, Ministry of Electronics and Information Technology, Government of India. He worked as head of the test and calibration laboratory and obtained ISO/IEC 17025 accreditation for the calibration laboratory in 2002. He also worked as head of the IT section. He is currently working as head of Quality Assurance and Training. He was instrumental in setting up the Quality System of ETDC as per ISO/IEC 17025.

Since 1998, he has given training courses, workshops, and seminars on Calibration Techniques, Measurement Uncertainty, Laboratory Management per ISO/IEC 17025, Certified Calibration Professional, and Statistical Process Control (SPC), conducted by ETDC, Pune.

He has published a few papers; including 'Guard-banding practices' in AdMET India 2012 and another, 'Proposed Guidelines for the Selection of Trapezoidal and Triangular Distributions for an Uncertainty Evaluation,' published in 'NCSL International Measure,' March 2013.

Currently working as Scientist 'F' and Director at the Electronics Test & Development Centre (ETDC), STQC Directorate, Ministry of Electronics and Information Technology, Agriculture College Campus, Shivajinagar, Pune, Maharashtra, 411005 (India).

List of abbreviations

ADC	Analog to digital converter
BIPM	International Bureau of Weights and Measures (Bureau International des Poids et Mesures)
CIPM	International Committee for Weights and Measures (Comité International des Poids et Mesures).
CL	Confidence level
CLT	Central limit theorem
CMC	Calibration measurement capability
CRM	Certified reference material
DMAIC	Define–measure–analyze–improve–control
DMM	Digital multimeter
DoF	Degrees of freedom
DUC	Device under calibration
DUT	Device under test
ESDM	Experimental standard deviation of the mean
FAR	False accept risk
FRR	False reject risk
GB	Guard band
GRR	Gauge R&R
GUF	GUM Uncertainty Framework
GUM	Guide to the expression of uncertainty in measurement
ILC	Interlaboratory comparison
LPU	Law of propagation of uncertainty
MCM	Monte Carlo method
MPU	Measurement process uncertainty
MSA	Measurement system analysis
NMI	National metrological institutes
PDF	Probability density function
PFA	Probability of false accept
PT	Proficiency testing
R&R	Repeatability and reproducibility
RP	Recommended practice
RSS	Root sum square
SL	Specification limit
SPC	Statistical process control
TAR	Test accuracy ratio
TL	Tolerance limit
TME	Test and measuring equipment
TUR	Test uncertainty ratio
VIM	International vocabulary of metrology

List of symbols

a	the half-width of a rectangular distribution of possible values of an input quantity X_i: $a = \frac{a_+ - a_-}{2}$
a_+	the upper bound, or upper limit, of input quantity X_i
a_-	the lower bound, or lower limit, of input quantity X_i
c_i	partial derivative or sensitivity coefficient: $c_i \equiv \frac{\partial f}{\partial x_i}$
$E[Y]$	Expectation of output quantity Y
f	the functional relationship between the measurand Y and input quantities X_i; also between the output estimate y and input estimates x_i
$\partial f/\partial x_i$	the partial derivative with respect to an input quantity X_i of the functional relationship f between the measurand Y and the input quantities X_i, evaluated at x_i for the X_i
k	a coverage factor used to calculate expanded uncertainty $U = ku_c(y)$
k_p	a coverage factor used to calculate expanded uncertainty $U_p = k_p u_c(y)$ at a level of confidence p
n	the number of repeated observations (used for sample size)
N	the number of input quantities X_i on which the measurand Y depends (used for population size)
p	probability; level of confidence
q	a randomly varying quantity described by a probability distribution
$\bar{q}$	the arithmetic mean or average of n independent repeated observations of randomly varying quantity q
$r(x_i, x_j)$	an estimated correlation coefficient associated with input estimates x_i and x_j $$r(x_i, x_j) = \frac{u(x_i, x_j)}{u(x_i)u(x_j)}$$
s	the standard deviation of the sample data
$s(\bar{x}), s(\bar{z})$	the experimental standard deviation of the sample mean of variable x or z
s_p	the pooled experimental standard deviation
$s(\overline{X_i})$	the experimental standard deviation of input mean $\overline{X_i}$
$t_p(v)$	t-factor from the t-distribution for v degrees of freedom corresponding to a given probability p
$t_p(\nu_{\text{eff}})$	The t-factor from the t-distribution for the ν_{eff} degrees of freedom corresponding to a given probability p, used to calculate the expanded uncertainty U_p
$u^2(x_i)$	the estimated variance associated with input estimate x_i
$u(x_i)$	the standard uncertainty of input estimate x_i
$u(x_i, x_j)$	the estimated covariance associated with two input estimates x_i and x_j
$u_c^2(y)$	the combined variance associated with output estimate y
$u_c(y)$	the combined standard uncertainty of output estimate y,
$u_i(y)$	a component of the combined standard uncertainty $u_c(y)$ of output estimate y generated by the standard uncertainty of input estimate xi: $u_i(y) \equiv ci\ u(x_i)$
$u(x_i)/\lvert x_i \rvert$	the relative standard uncertainty of input estimate x_i
$u_c(y)/\lvert y \rvert$	the relative combined standard uncertainty of output estimate y
U	the expanded uncertainty of output estimate y: $U = ku_c(y)$
U_p	the expanded uncertainty of output estimate y: $Up = k_p u_c(y)$
x_i	an estimate of input quantity X_i

X_i	the ith input quantity on which measurand Y depends
$\overline{X}_i$	an estimate of the value of input quantity X_i
y	an estimate of measurand Y *or* the result of a measurement
Y	a measurand
ν	degrees of freedom
ν_i	degrees of freedom of the standard uncertainty $u(x_i)$ of input estimate x_i
ν_{eff}	effective degrees of freedom
$\overline{x}$	the arithmetic mean or average of n independent repeated observations of randomly varying quantity x
z	a statistic, a random variable, normalized error
$\overline{z}$	the arithmetic mean or average of z
α	significance level, producer's risk
σ^2	the variance of a probability distribution of a randomly varying quantity
$\sigma(\overline{q})$	the standard deviation of $\overline{q}$
$\sigma[s(\overline{q})]$	the standard deviation of an experimental standard deviation $s(\overline{q})$ of $\overline{q}$
μ	a population mean or average

How to read this book

To be frank, this book is written in such a manner that practicing metrologists can open any chapter and start reading. However, for beginners, it would be best to start with part A and follow through in sequence. The parts and chapters are arranged in an ascending order of conceptual development and understanding. However, a few chapters provide cross-references to other chapters which may be read by practicing metrologists in continuation of that chapter for ease of understanding.

Here is a brief synopsis of parts A through F.

Part A is titled '**Understanding fundamentals**' and discusses essential concepts related to the evaluation of measurement uncertainty. It familiarizes the reader with various terminologies and would certainly be recommended for beginners to commence there.

Part B is '**Dealing with distributions.**' This part addresses the process of understanding various distributions, which is very important for arriving at an optimal evaluation of uncertainty.

Part C is about '**Sample size and analysis.**' Sampling, sample preparation, the homogeneity of samples, etc. are key factors that contribute to uncertainty evaluation. The determination of sample size is an issue commonly faced by metrologists and is considered in this part.

Part D is '**Decoding degrees of freedom.**' 'Degrees of freedom' is again a difficult term for metrologists to grasp, and the three chapters of this part try to give the reader an insight into this concept.

Part E is titled '**Some contiguous concepts.**' In the process of evaluating uncertainty, there are many associated terms and concepts which should be properly understood. This part addresses some of those.

Part F is '**Delving a little deeper.**' This part is meant for experienced or seasoned metrologists. It considers some advanced concepts and also discusses some important notes in the GUM that are essential for a good understanding of the subject.

NOTE:

- The foreword of ISO/IEC Guide 98-3:2008 mentions that 'This first edition of ISO/IEC Guide 98-3 cancels and replaces the *Guide to the Expression of Uncertainty in Measurement (GUM)*, BIPM, IEC, IFCC, ISO, IUPAC, IUPAP, OIML, 1993, corrected and reprinted in 1995.'
- ISO/IEC Guide 98-3:2008 is a reissue of the 1995 version of *Guide to the Expression of Uncertainty in Measurement* (GUM), also called 'JCGM 100:2008,' with minor corrections. Thus, wherever text in this book refers to the GUM, it also refers to ISO/IEC Guide 98-3:2008.
- Some documents (standards, guides, etc.) have been updated but reference to earlier versions or editions is made for illustration and a better understanding of the concept. Reference to both versions/editions is given at a few places.

- Unless specifically mentioned otherwise, the word 'uncertainty' in the book means 'measurement uncertainty'.
- Some documents mentioned in the 'References' section may not be available or may have been withheld/withdrawn. Reference to such documents is given for exemplification and discernment of the subject. Readers are advised to refer only to the current versions/editions of documents in practice.
- The short term 'combined uncertainty' is used in many places, which may be treated as 'combined standard uncertainty'. The former term is used in Recommendation INC-1 of the Working Group of CIPM while the latter term is adopted by GUM.

Part A

Understanding the fundamentals of
measurement uncertainty

IOP Publishing

A Practical Handbook on Measurement Uncertainty
FAQs and fundamentals for metrologists
Swanand Rishi

Chapter 1

Using correct terminology

Abstract: In any scientific work, terms and definitions have specific connotations and are carefully crafted. Thus, it is imperative to use correct terminology in order to convey what one wants to say.

Finally there are simple ideas of which no definition can be given; there are also axioms or postulates, or in a word primary principles, which cannot be proved and have no need of proof.

—Gottfried Leibniz

Metrology (as defined in VIM 2.2 [1]) is the 'science of measurement and its application.' Obviously, like all other subjects, it has its own terminology. It is expected that the terms should be understood in the correct sense and used in the right context. As metrology is fundamental to the development of science and technology, its terminology has significant importance. Thus, it is pertinent that all its terms are used carefully and properly.

We will see a few terms that are found to be used wrongly, unknowingly, or otherwise. Most terms can be found in the VIM (Vocabulary of International Metrology; also called JGCM 200:2012 [1]).

1.1 Quantity and quantity value

Quantity (VIM 1.1 [1]): It is 'property of a phenomenon, body, or substance, where the property has a magnitude that can be expressed as a number and a reference.' Thus, any quantity has a numerical value followed by a reference; typically, a unit, although it may be a measurement procedure or a reference material. For example, length and Rockwell Hardness are quantities because, when they express a measurement, they have values associated with them followed by units (meter and HRC, respectively. The first example given uses a unit while the second uses a procedure— Rockwell C).

doi:10.1088/978-0-7503-6462-1ch1

Quantity, being a property or an attribute, needs to be expressed in adequate detail. Otherwise, it is just a description of a property and does not contain value. (However, the value when expressed should have magnitude and reference.) As examples, length as a quantity can be described as the 'length of the steel rod at 20 °C' and temperature as a quantity can be described as the 'temperature of a given sample after storage for 1 h.'

Implicit in the meaning of quantity is that it should be measurable. Therefore, the GUM B.2.1 [2] defines a 'measurable' quantity as the 'attribute of a phenomenon, body or substance that may be distinguished qualitatively and determined quantitatively.'

Quantity Value (VIM 1.19 [1]): It is defined as 'number and reference together expressing magnitude of a quantity.' For example, 'length of the steel rod: 10.25 cm' or 'temperature of the given sample: 50.36 °C.' Thus, it is value of a property that is expressed in terms of a magnitude and a reference. 'Quantity' and 'quantity value' are different terms and have specific meanings. In general, all measurands have quantity values. In the GUM, the terms 'result of measurement' and 'estimate of the value of the measurand' or just 'estimate of the measurand' are used for 'measured quantity value'.

A fact one should know

From the above definitions length, power, and luminous intensity (for example) are quantities; but (again for example) accuracy, precision, and error are not. For we do not 'measure' accuracy or precision, rather we calculate their values from the measured data of input estimates. Thus, one should not say 'precision of the inductor is 0.05 mH,' rather it would be correct to say the 'quantity value of the precision of the inductor is 0.05 mH.'

The following examples and notes under 'quantity value,' reproduced from VIM [1], would make excellent reading and enhance one's understanding of 'quantity value.'

EXAMPLE 1: Length of a given rod: 5.34 m or 534 cm

EXAMPLE 2: Mass of a given body: 0.152 kg or 152 g

EXAMPLE 3: Curvature of a given arc: 112 m^{-1}

EXAMPLE 4: Celsius temperature of a given sample: −5 °C

EXAMPLE 5: Electric impedance of a given circuit element at a given frequency, where j is the imaginary unit: (7 + 3j) Ω

EXAMPLE 6: Refractive index of a given sample of glass: 1.32

EXAMPLE 7: Rockwell C hardness of a given sample: 43.5 HRC

EXAMPLE 8: Mass fraction of cadmium in a given sample of copper: 3 μg kg^{-1} or 3×10^{-9}

EXAMPLE 9: Molality of Pb^{2+} in a given sample of water: 1.76 μmol kg^{-1}

EXAMPLE 10: Arbitrary amount-of-substance concentration of lutropin in a given sample of human blood plasma (WHO International Standard 80/552 used as a calibrator): 5.0 IU l^{-1}, where 'IU' stands for 'WHO International Unit'

NOTE 1: According to the type of reference, a quantity value is either
— a product of a number and a measurement unit (see examples 1, 2, 3, 4, 5, 8, and 9); the measurement unit 'one' is generally not indicated for quantities of dimension one (see examples 6 and 8), or
    — a number and a reference to a measurement procedure (see example 7), or
    — a number and a reference material (see example 10).
NOTE 2: The number can be complex (see example 5).
NOTE 3: A quantity value can be presented in more than one way (see examples 1, 2, and 8).
NOTE 4: In the case of vector or tensor quantities, each component has a quantity value. EXAMPLE: The force acting on a given particle, e.g. in Cartesian components $(Fx; Fy; Fz) = (-31.5; 43.2; 17.0)$ N.

1.2 The measurand and parameter

The measurand (VIM 2.3 [1]): It is a 'quantity intended to be measured,' while GUM B.2.9, the second edition of VIM, and IEC 60050–300:2001 define it as a 'particular quantity subject to measurement.' It is a well-defined physical quantity that can be characterized by an essentially unique value. Thus, as in case of 'quantity,' the measurand also needs to be defined in adequate detail with respect to the known influence factors. The objective of measurement is to find value of the measurand; hence the method and procedure of measurement need to be elaborated. For example, to measure power dissipated in a resistor at a certain temperature, the applied voltage, the temperature itself, the temperature coefficient of the resistor (α), and R_0 should be defined. The measurement system may affect the measurand (phenomenon, body, or substance), in which case a correction shall be applied.

As clarified in the GUM [2], the measurand cannot be specified by a value but only by a description of a quantity. In practice, the definition of the measurand is also dictated by the required accuracy of measurement and the physical states and conditions, so that for all practical purposes its value is essentially unique. Taking the above example, if the accuracy of measurement is, say, 10 ppm, the nonlinearity of α or a nonuniform temperature gradient, or even barometric pressure and air density may have to be defined. For low accuracies of, say, 0.01% these additional factors may not be required to be defined. The GUM gives an example: 'the velocity of sound in dry air of composition (mole fraction) $N_2 = 0.780\ 8$, $O_2 = 0.209\ 5$, $Ar = 0.009\ 35$, and $CO_2 = 0.000\ 35$ at temperature $T = 273.15$ K and pressure $p = 101\ 325$ Pa.'

The result of measuring a measurand is given by its best estimate and the associated uncertainty; $Y = y \pm U$. As the word 'true' is redundant in the GUM, it advocates use of term 'value of the measurand' and not 'true value of the measurand.' The GUM uses the expression 'value of the measurand' with the above consideration.

As in case of quantity, the terms accuracy, precision, bias, etc. are not measurands for the same reason.

A parameter (GUM C.2.7 [2]): It is 'a quantity used in describing the probability distribution of a random variable.' Often, this term is erroneously used to describe a measurand. For example, 'calibration of temperature parameter' or 'salinity of soil is the parameter under test.' Such terminology should be avoided because, according to the definition, the term 'parameter' relates to a variable that can be described by a distribution. Thus, it is used in the GUM for terms such as uncertainty (its definition starts with 'parameter'), standard deviation, degrees of freedom, etc. Frequently, it appears in the description of concepts related to population, such as 'population parameters like mean or standard deviation' (in clause C.2.23 of GUM [2]) or '... statistical parameters characterising...' (in clause E.4.4 of GUM [2]).

Apart from describing variables/factors of a statistical nature, this term is also used to describe influencing factors or other variables affecting the measurement of a measurand. This is evident from clause 3.3.2(h) of GUM [2], which gives some examples of sources of uncertainty where it states '...and other parameters obtained from external sources....' Thus, if we are measuring flow, the 'flow' is the 'measurand' and not the 'parameter.' The parameters are those quantities that affect the measurement of the flow, depending upon the method. For example, in the static gravimetric method, the temperature, timer actuation, diverter time, buoyancy correction, scale calibration and drift, evaporation, etc. are likely to be the parameters. Similarly, when calibrating a capacitor, the capacitance is the measurand and, depending upon level of accuracy, the frequency, test voltage, hysteresis, and temperature could be the parameters.

In practice, this distinction is usually overlooked and the two words are used interchangeably. This practice should be avoided in principle.

In IS 7920 (Part 1):2012 [3], 'parameter' is defined as the 'index of a family of distributions.' By 'index,' it means the defining aspect or characteristic of the distribution, e.g. the mean or standard deviation. The parameters (of a distribution) are also those that define the location, scale, or shape of the distribution.

1.3 Accuracy and precision

Accuracy (VIM 2.13 [1]): It is the 'closeness of agreement between a measured quantity value and a true quantity value of a measurand.' As per the GUM, it is the 'closeness of the agreement between the result of a measurement and a true value of the measurand.' In both definitions, accuracy is expressed with respect to some reference (true) value. Accuracy is a *qualitative* term and is determined by trueness and precision, among other factors. Thus, it should not be confused with trueness, which is the 'closeness of agreement between the average of an infinite number of replicate measured quantity values and a reference quantity value.' Accuracy (like trueness) is *not* a quantity value and hence should not be accompanied by a numerical value. For example, it is wrong to say that 'the accuracy of a standard inductor is 0.1 mH.' The right way is: 'the numerical quantity value of accuracy of a standard inductor is 0.1 mH.' A measurement is supposed to be more accurate when it has smaller measurement error.

Measurement Precision (VIM 2.15 [1]): It is the 'closeness of agreement between indications or measured quantity values obtained by replicate measurements on the same or similar objects under specified conditions.' Thus, precision is only an indicator of the spread or dispersion of the data and has nothing to do with the reference value. The GUM does not define it separately but defines a similar term 'repeatability' as the 'closeness of the agreement between the results of successive measurements of the same measurand carried out under the same conditions of measurement.' Thus, precision just shows how close the readings are to each other and the standard deviation is considered to be the measure of their closeness.

Often, there is a query whether accuracy is more important or precision. If we look at measurement as a controlled process (and it is so), a process is said to be under statistical control when it is influenced by nonassignable causes or random effects only. Measurement is more precise when its variability is less (i.e. it is more consistent), and this is the result of good manufacturing practices that include design, material, and process. Given that the material and process are well established, higher precision can be achieved by appropriate design. Thus, precision is a function of design and can be built into the product. Further, a good design provides for adjustment or 'trimming' in case readings are out of tolerance due to some other factors. A poor precision is undesirable, as it is an indication of unpredictable device behavior. (Like, a consistent or predictable enemy is easier to deal with than an unpredictable friend!) Since precision (in addition to bias) is part of accuracy, high precision necessarily results in a high accuracy, for a given bias.

Thus, the two terms 'accuracy' and 'precision' are neither synonymous nor comparable owing to their definitions and interdependence. The term 'precision', like accuracy, is also *qualitative* in nature and should not be used to describe a quantity value, as explained for accuracy.

1.4 Standard uncertainty and uncertainty contribution

Standard uncertainty (or standard measurement uncertainty) (VIM 2.30 [1]): It is defined as 'measurement uncertainty expressed as a standard deviation.' The GUM defines it as the 'uncertainty of the result of a measurement expressed as a standard deviation.' So, it is obtained by taking the standard deviation of the error of the input estimate. (The standard deviation of the error should not to be called the 'standard error.') The standard uncertainty should not be confused with the (final) uncertainty contribution of the component under evaluation. There is no specific term in the GUM (or in the VIM) for the final uncertainty contribution of the component under evaluation, but it is shown by the symbol $u_i\,(y)$. It appears in GUM [2] in equation 11(a) for the first time as a part of the equation for the combined variance of an output estimate and is explained in 11(b) by the equation

$$u_i\,(y) \cong |c_i|\,u\,(x_i)$$

$$\cong \text{sensitivity coefficient} \times \text{standard uncertainty of input estimate.}$$

It is described in Annexure J of the GUM as a 'component of combined standard uncertainty $u_c(y)$ of output estimate y generated by the standard uncertainty of input estimate x_i.' UKAS M3003:2007 [4] uses this term in various places, but mentions it as 'standard' uncertainty (in table under cl. 3.40), and explains by a footnote that it is 'scaled in accordance with the effect of the input quantity.' ('Scaled' by multiplying the standard uncertainty by the sensitivity coefficient.) EUROLAB Technical Report No. 1/2006 and the Indian Guide NABL 141 [5] have used an apt term *'uncertainty contribution'* for $u_i(y)$. As there is no specific term in the GUM for the final uncertainty of an *input* estimate, it is worth describing it as *'uncertainty contribution.'*

Uncertainty Contribution: NABL 141 has described it as 'the contribution to the standard uncertainty associated with the output estimate y resulting from the standard uncertainty associated with the input estimate x_i.' Technically, it is same as what the GUM denotes by $u_i(y)$. Thus, the uncertainty contribution of an input estimate is obtained by multiplying the standard uncertainty of that estimate by the corresponding sensitivity coefficient.

A fact one should know

It is the uncertainty contribution $u_i(y)$ and not the standard uncertainty $u(x_i)$ that is taken to calculate the combined uncertainty. In principle, the standard uncertainty is not the uncertainty contribution of that component. Although in the majority of cases the sensitivity coefficient is one, that gives an uncertainty contribution equal to standard uncertainty, this fine distinction should be borne in mind.

1.5 Error, uncertainty, and expanded uncertainty

Like accuracy and precision, error and uncertainty are terms that are often confused with each other. We discuss error in chapter 5, 'Various error terms and bias.' Here is a quick recap of the two terms.

Error (VIM 2.16 [1]): It is the 'measured quantity value minus a reference quantity value.' A reference quantity value is either a 'true value' or a 'conventional true value.' In the former case, it is unknown, and hence the error is also unknown. In the latter case, which is very rare and only valid in very limited cases, it is known, and hence the error is also known. In general, the former case prevails and thus error is an idealized concept and cannot be known exactly. It arises because no measurement is perfect. We can just try to minimize error by making a sufficiently large number of measurements and minimizing the known influence factors and correcting them. But in spite of all this, error always remains. Sometimes error may appear to be zero— with or without correction—but it can then be deemed to be zero vis-à-vis the then-prevailing state of our knowledge. It is, in principle, unknown and unknowable— forever!

The concept of error revolves around the reference value and is treated mathematically. The relation between error and uncertainty is like that of cause and effect—the error is the cause that results in uncertainty. As error cannot be eliminated completely, uncertainty always remains. (This applies in most cases. The instances in which the measurand is supposed to be exactly known, resulting in zero uncertainty, are dealt with in chapter 5, 'Various error terms and bias.')

Uncertainty (VIM 2.26 [1]): It is a 'non-negative parameter characterizing the dispersion of the quantity values being attributed to a measurand, based on the information used.' We are more familiar with the definition given by GUM [2] (B.2.18) that defines it as a 'parameter, associated with the result of a measurement that characterizes the dispersion of the values that could reasonably be attributed to the measurand.' Both definitions basically convey the same meaning. The parameter is mostly the standard deviation (when estimated), or the half-width of an interval at a stated confidence level (when borrowed from other data or assumed).

The concept of uncertainty revolves around the result of measurement which is believed to be the best estimate of the measurand (i.e. expectation or mean) and treated statistically.

Despite this radical approach, the GUM points out that the new definition is consistent with old concepts of uncertainty as 'a measure of possible error in estimated value of measurand' and 'an estimate characterizing the range of values within which the true value of a measurand lies.'

It should also be noted that the terms 'uncertainty' and 'expanded uncertainty' are not synonymous! The former is a generic term and various adjectives modify it for the intended definition. For example, 'combined uncertainty,' 'standard uncertainty,' and 'expanded uncertainty' are the modifiers that define different concepts related to uncertainty. The confusion occurs because 'expanded uncertainty' is usually the *final* uncertainty of a measurement.

Expanded uncertainty (VIM 2.35 [1]): It is the 'product of a combined standard measurement uncertainty and a factor larger than the number one.' Here, the 'factor' is the coverage factor k. (The reason that k should be >1 is explained in the box below.) The expanded uncertainty is termed the '*overall uncertainty*' in paragraph 5 of Recommendation INC-1 (1980). In GUM [2] (clause 2.3.15), expanded uncertainty is defined as a 'quantity defining an interval about the result of a measurement that may be expected to encompass a large fraction of the distribution of values that could reasonably be attributed to the measurand.' Although the definitions of both uncertainty and expanded uncertainty given in the GUM convey essentially similar meanings, the terms should not be considered synonymous.

A fact one should know

It is worth noting that the VIM definition of expanded uncertainty points out that $k > 1$, whereas the GUM states in the accompanying note that it typically lies between 2 and 3. Recommendation 1 (INC 1–1986), recommends that the combined uncertainty of Type A and Type B components shall be given in terms of *one standard deviation*. The combined standard uncertainty is believed to be a standard Normal distribution characterized by a mean $\mu = 0$ and a standard deviation $\sigma = 1$. As we want an '*expanded*' uncertainty which is obtained from the combined standard uncertainty, the latter must be multiplied by a factor (the coverage factor, k, in our case) which is *more* than one (and not *more than or equal to one*). This also means that it has to be estimated at a level of confidence of more than 68.27%, corresponding to effective degrees of freedom of $\nu_{\text{eff}} \approx \infty$. This is usually done at an approximately 95% level of confidence.

By the way, the symbol for the coverage factor should be written as 'k' (in *italics*), and not roman 'k' or 'K' ('K' being the symbol for 'Kelvin'—an SI unit for thermodynamic temperature). Metrology has a close connection with the SI. In the SI, 'k' is a symbol used as a prefix for 'kilo,' (except in case of the unit of mass, where it is not a prefix but has been retained for historical reasons). All metrologists should be conversant with the correct use of SI units, symbols, and prefixes.

References

[1] VIM 2012 *International Vocabulary of Metrology – Basic and General Concepts and Associated Terms* 3rd edn (Sèvres: BIPM) JCGM 200:2012

[2] JCGM 100:2008 *(GUM) Evaluation of Measurement Data—Guide to the Expression of Uncertainty in Measurement* 1st edn (BIPM, IEC, IFCC, ISO, IUPAC, IUPAP, OIML—International Organization for Standardization)

[3] IS 7920 (Part 1): 2012 (R2017) *Statistics—Vocabulary and Symbols Part 1: General Statistical Terms and Terms Used in Probability (Third Revision)* (New Delhi: Bureau of Indian Standards)

[4] UKAS M3003 2007 (revised/reviewed and confirmed in 2022) *The Expression of Uncertainty and Confidence in Measurement* (Staines-upon-Thames: United Kingdom Accreditation Service)

[5] NABL 141 2020 *Guidelines for Estimation and Expression of Uncertainty in Measurement* (Gurugram: NABL India) Issue 4

Chapter 2

Why do we need to evaluate uncertainty?

Abstract: In spite of one's best efforts, no measurement is perfect. The result of a measurement is only an approximation or estimate of the value of the measurand and is hence incomplete without a statement of the accompanying uncertainty. Further, some measurements may have to be qualified by a statement of conformity or may require the application of a decision rule as per ISO/IEC 17025 [1], hence the need to evaluate uncertainty.

> *You cannot step twice into the same river.*
>
> —Heraclitus

> *An approximate answer to the right problem is worth a good deal more than an exact answer to an approximate problem.*
>
> —John Tukey

The progress of humanity owes a lot to all types of measurements, particularly scientific measurements. Measurements are performed in order to gather information about an object of interest. They are essentially investigations performed to quantify some parameter or attribute of the object. In metrology, we call such a parameter or attribute a 'measurand'—a particular quantity subject to measurement.

The first step in measurement is to define the quantity. This is followed by the realization, representation, and dissemination of the quantity. Each of the above steps is subject to our current knowledge, skill, experience, interpretation, etc. Based on gathered experience and improved technology, the definitions of measurands are changed, thus affecting subsequent steps. The famous DMAIC cycle; 'Define, Measure, Analyze, Improve, and Control,' does not end at 'control.' It is not an open-ended scheme of things. It is a '*cycle*,' and once sustainable 'control' is attained, the 'definition' is changed (or rather needs to be changed), thereby

doi:10.1088/978-0-7503-6462-1ch2

changing the measurement scenario. It is worth noting that the definitions of all seven base SI units were revised with effect from 20th May, 2019, although the changes were not very significant in a few cases.

The result of measurement broadly depends upon the following factors:

1. The object itself
2. The measurement system
3. The measurement method/procedure
4. The test environment
5. The operator

We know that no measurement gives the same result when we take several measurements, even when we keep all the above factors same and have excellent control. (Of course, with poor resolution, one may get the same result, but this does not mean that the measurement is perfect. We are also ignoring cases in which the same results are 'obtained' through personal bias, manipulated data, blunders, etc. Irrespective of best practices, no measurement is perfect.) Hence, we are never sure whether we will get the same result next time. This casts some doubt on the result of the measurement.

The result of a measurement is obtained by taking the average (mean) of several (but a limited number of) readings, called 'sample data' and it is assumed to represent the 'population' of readings. Thus, the mean is supposed to be the best *estimate* (but an '*estimate*' because it is based on limited data) of the quantity being measured. Being a single value, it is a 'point' estimate, and will almost certainly be different for another sample set. So, it raises 'doubt' about the current result. In order to know by how much this result may vary, statisticians have developed the concept of the 'interval estimate.' This is an interval around the result of measurement, which is 'believed' to contain an infinite number of possible values of the measurand, if repeat measurements are taken under same conditions as before. This interval is called the 'confidence interval estimate' or the 'uncertainty' of measurement. (It is the expanded uncertainty, to be precise.) It is estimated based on the standard deviation of the dispersion of values in the sample data (Type A evaluation) and other factors that are known to contribute to the result (Type B evaluation).

The dilemma is that we cannot be 100% sure about this interval, as it is basically an *estimation* that is based on limited knowledge and obtained under several practical constraints. Therefore, using this interval estimate, we cannot predict the population characteristic of interest with 100% certainty. This difficulty is surmounted by setting the confidence level to less than 100%. A typical level of confidence is 95%, which is recommended in the GUM [2] and almost all other guides. This means that if the same measurement is repeated several times under similar conditions, we can expect the earlier result to lie in the given interval 95% of the time (19 times in 20).

A fact one should know

The meaning of 'confidence interval' needs elaboration. The confidence interval corresponds to the probability of obtaining a previously estimated *mean* value (the result of measurement) that falls in that interval. This estimated interval applies to the *mean* value. It does not imply that probability of getting *any* value within that interval is 95%.

This is how measurement uncertainty comes into play. As explained in JCGM 101 (JCGM-Joint Committee for Guides in Metrology) [3], measurement uncertainty reflects our incomplete knowledge of the measurand. It is a measure of how well one *believes* one knows the essentially unique 'true' value of the measurand, as it is not possible to state how well the essentially unique true value of the measurand is *known*. The notion of 'belief' is an important one, since it moves metrology into a realm in which the results of measurements need to be considered and quantified in terms of probabilities that express degrees of belief.

The uncertainty of measurement also helps us to compare results and check the capabilities of different laboratories. This is particularly important when the quantity value of the estimated result (mean) is same (for the given resolution).

For example, let the results of three readings of a measurand for laboratory 'A' be 105, 100, and 95 units against a reference of 100 units. Let the corresponding results for laboratory 'B' be 101, 100, and 99 units. The mean of both is 100 units, which is the best estimate. Hence both are equally 'accurate' or have the same 'error' and thus we cannot judge or evaluate their capabilities from their 'means.' It is also quite possible that two laboratories or two operators can get same result or the same error using different methods or apparatus. Thus 'equal accuracy' or 'same error' is inadequate as an indicator of a laboratory's capability. This limitation can be overcome by comparing their uncertainties. This is another reason that uncertainty comes into the picture as a decisive and conclusive measure of measurement capability.

In the above example, laboratory 'B' has lower *dispersion* (spread of values) and hence lower uncertainty for the measured set of values. Therefore, it has a definitive edge over laboratory 'A' as regards their measurement capabilities.

Thus, we need to evaluate uncertainty because

1. Evaluation of uncertainty is a prerequisite to get accreditation for laboratories as per clause 7.6.1 of ISO/IEC 17025:2017 [1], which requires that 'Laboratories shall identify the contributions to measurement uncertainty. When evaluating measurement uncertainty, all contributions that are of significance, including those arising from sampling, shall be taken into account using appropriate methods of analysis.'

2. It quantifies quality of measurement.
3. We can understand the major contributions to variability and can attempt to eliminate or minimize them.
4. We can meaningfully compare results from the same as well as different sources. This avoids unnecessary repetition of the same test methods if the results are close.
5. In certain situations, e.g. in legal metrology, the purpose of measurement or calibration is to check conformity to specifications. The uncertainty of measurement in such cases is usually negligible and not worth evaluating. But without evaluation, it remains elusive and will need formal evaluation if challenged. Uncertainty will then be a decisive factor.
6. It helps the customer to interpret data, particularly in conformity decisions.
7. It helps in method validation, bias, and trueness studies.

A fact one should know

There is one exception to reason (1) above. Note 1 to Clause 7.6 of ISO/IEC 17025:2017 says '*In those cases where a well-recognized test method specifies limits to the values of the major sources of uncertainty of measurement and specifies the form of presentation of calculated results, the laboratory is considered to have satisfied clause 7.6.3 by following the test method and reporting instructions.*' Note 2 of the same standard also excludes evaluation of uncertainty for test laboratories under specific conditions. See chapter 10 'Some uncommon uncertainties' and chapter 34 'Alternative approaches in uncertainty evaluation'

A point to ponder

Let us take a detour to the evolution of the 'uncertainty principle' concept that is one of the foundations of quantum mechanics. The idea and use of uncertainty in metrology dates back to the 1950s. However, in theoretical physics, the concept was established way back in 1927 when Heisenberg announced his famous uncertainty principle in a paper. He received the Nobel prize for Physics in 1932 'for the creation of quantum mechanics,' to which the uncertainty principle made a seminal contribution. The uncertainty principle states that '*It is impossible to 'simultaneously' measure both position and momentum of a particle with infinite precision.*' Thus, when one tries to measure position (x) with very high precision, one ends up measuring momentum (p) with poor precision, as described by the equation

$$\Delta x \times \Delta p = h,$$

where Δ stands for uncertainty (in terms of standard deviation) and h is Planck's constant (6.626×10^{-34} J s, with standard uncertainty of 29×10^{-42} J s) In short, Planck's constant is the limit due to which the '*point estimates*' of position and momentum cannot be *simultaneously* determined exactly. However, their '*interval estimate*' can be determined with some *probability*!

It may be noted that due to limits on the experimental measurement errors, the principle has nothing to do with practical measurements and was postulated as a law of nature. (Effective from 20th May 2019, the Planck's constant has the exact value of $6.62607015 \times 10^{-34}$ J s).

References

[1] ISO/IEC 17025:2017 *General Requirements for the Competence of Testing and Calibration Laboratories* (Geneva: ISO)

[2] JCGM 100:2008 *(GUM) Evaluation of Measurement Data—Guide to the Expression of Uncertainty in Measurement* 1st edn ((BIPM, IEC, IFCC, ISO, IUPAC, IUPAP, OIML—International Organization for Standardization)

[3] JCGM 101:2008 *Evaluation of Measurement Data—Supplement 1 to the 'Guide to the expression of uncertainty in measurement'—Propagation of Distributions Using a Monte Carlo Method* 1st edn (Sèvres: BIPM)

IOP Publishing

A Practical Handbook on Measurement Uncertainty
FAQs and fundamentals for metrologists
Swanand Rishi

Chapter 3

Should we talk about the 'evaluation' or 'calculation' of uncertainty?

Abstract: In any scientific work, terms and definitions have specific meanings and are carefully crafted. There are distinct meanings to the words 'evaluation,' 'estimation,' and 'calculation.' The common mistake of talking about the *calculation* or *estimation* of uncertainty should be avoided.

> *While the individual man is an insoluble puzzle, in the aggregate he becomes a mathematical certainty. You can, for example, never foretell what any one man will be up to, but you can say with precision what an average number will be up to. Individuals vary, but percentages remain constant. So says the statistician.*
>
> —Arthur Conan Doyle

During seminars on measurement uncertainty and on other occasions, participants generally use the word 'calculation' while talking about uncertainty. The word 'calculation' is used habitually, missing the subtle difference between 'calculation' and 'evaluation.' 'Calculation' is a generic word, and although it may comprise estimation or computation, it is inappropriate to use it as a substitute for 'evaluation.'

The word 'calculation' means any mathematical operation. For example, we solve equations involving trigonometric, integral, or derivative functions, which by and large are calculations. There are clear-cut rules and formulae to be used when we calculate something. In uncertainty evaluation, we also *calculate* the mean of the sample data. There is a fixed formula for the calculation of the mean. But in uncertainty evaluation, 'calculation' *per se* ends there, and we enter into the realm of '*evaluation*' after the mean is established.

Clause C.2.24 of the GUM [1] defines 'estimation' as 'the operation of assigning, from the observations in a *sample*, numerical values to the parameters of a distribution *chosen* as the statistical model of the population from which this sample is taken.'

Estimation is a specific word meaning an approximation, assessment, valuation, or guess, based on a randomly drawn sample that is assumed to represent the population.

As all metrologists know, in Type A evaluation, we calculate the mean of the sample data that are very few in number and assume that it closely represents the population mean. It is stated in NOTE 1 under clause 4.2.3 of the GUM that 'the number of observations n should be large enough to ensure that q provides a reliable estimate of the expectation μ_q of the random variable q.' But, for practical reasons, the sample data collected are quite few in number, and we *assume* that the mean of the sample data closely represents the population mean. In measurements of small sample size, we deal with what is called the 'strong law of small numbers,' formulated by the British mathematician Richard Guy, which states that 'There aren't enough small numbers to meet the many demands made of them!' Thus, strictly speaking, because of the assumption, one can say that the mean is also an estimation!

Points to ponder
  ➢ Note that the term 'mean' is used generally when referring to a *population* parameter and the term 'average' when referring to the result of a calculation on the data obtained in a *sample*.
  ➢ A result obtained through estimation may be a single value (point estimate) or an interval estimate. So 'average' is a 'point estimate,' whereas 'uncertainty' is an 'interval estimate.'

However, let us gloss over this very fine detail, and move on to the Type B method of evaluation.

In Type B evaluation, we rely on data other than practical observations. Although the source and reliability of data may pose doubt, it is customary and acceptable to use this data with an *assumed* probability distribution. The probability is viewed as a measure of the *degree of belief* that an event will occur. As per the NOTE to clause 3.3.5 of the GUM, the Type B evaluation of an uncertainty component is usually based on a pool of *comparatively reliable* information. Furthermore, in certain cases, we may consider one or two additional factors based on experience and knowledge with *assumed* reliability. Thus, Type B evaluation is essentially based on *degree of belief*. Hence, wherever we use data based on probability, guess, degree of belief or assumptions, we 'estimate' rather than 'calculate' that quantity value or parameter.

A fact one should know

If you examine the GUM meticulously, it uses the words 'evaluation,' 'estimation,' or 'assessment' for uncertainty. It uses the word 'calculation' very sparingly (and not for uncertainty). For example, in clauses 5.1.3, 5.1.4, and G 4.2 it uses the verb 'calculate,' evidently because there is a straightforward function or formula for the *calculation* of those parameters or entity. The US guide NIST Technical Note 1297 in clause D.1.6 also clarifies this thus—'The word 'uncertainty,' by its very nature, implies that the uncertainty of the result of a measurement is an estimate and generally does not have well-defined limits.'

Another important reason for avoiding the use of word 'estimation' is the abolition of the concept of 'true' value in the GUM. This is because the 'true' value is the one which is supposed to be exactly and absolutely known (but practically unknowable) as compared to an 'estimated' value which is based on limited knowledge *and* some degree of belief.

Therefore, etymologically, the correct word to describe the process of 'working out' or establishing the uncertainty is 'evaluation' (or, with limited application, 'estimation') and *not* 'calculation.' As mentioned in clause 2.2.4 of the GUM '.....an uncertainty component is always *evaluated* using the same data and related information.' Note that even the GUM's title is '**Evaluation** of measurement data— Guide to the **expression** of uncertainty in measurement!

NOTE: Although the GUM uses the word 'evaluation' in a particular sense, many standards/guides use the word 'estimation' in their titles. For example,
   1. ISO 19036:2019: Microbiology of the food chain—Estimation of measurement uncertainty for quantitative determinations.
   2. ISO 14253-2 International Standard, 'Geometrical Product Specifications (GPS)— Inspection by measurement of workpieces and measuring instruments—Part 2: Guide to the Estimation of Uncertainty in GPS Measurement in Calibration of Measuring Equipment and in Product Verification,' 1999. [Revised 2011]

However, following the GUM, as the apex document in evaluation of measurement uncertainty, it is worth using the term 'evaluation' instead of 'estimation.'

Reference

[1] JCGM 100:2008 *(GUM) Evaluation of Measurement Data—Guide to the Expression of Uncertainty in Measurement* 1st edn (BIPM, IEC, IFCC, ISO, IUPAC, IUPAP, OIML— International Organization for Standardization)

Chapter 4

What uncertainty is not

Abstract: Many terms such as error, tolerance, accuracy etc. are mistakenly used to mean uncertainty. All scientific terms have specific meanings and should be used with the utmost care. The term 'uncertainty' is no exception, and should not be used to convey some convenient assumption.

> *There are three types of lies—lies, damn lies, and statistics.*
> —Benjamin Disraeli

We are familiar with the definition of uncertainty given by the GUM (B.2.18) [1], which defines it as a 'parameter, associated with the result of a measurement that characterizes the dispersion of the values that could reasonably be attributed to the measurand.' So, uncertainty is a range of values within which the best estimate (i.e. mean of sample set of readings) is expected to lie when the experiment is repeated under the same conditions. So, in a way, in spite of taking every precaution to maintain the 'integrity' of the measurement system, a pragmatic metrologist must have some doubt about the measured value. Uncertainty is the doubt about this measurement.

However, many other terms and concepts are misconceived or used as alternative terms for uncertainty, as we shall see below.

1. **Error is not uncertainty**

   Broadly speaking, errors can be classified as 'gross errors,' 'random errors' and 'systematic errors.' Gross errors arise mainly due to operator negligence, incorrect data recordings, or personal bias and can be overcome by proper supervision and training. Error, when properly expressed in terms of its standard deviation, is the uncertainty of that error component. This uncertainty, when multiplied by the sensitivity coefficient, is the uncertainty contribution of that error. An important distinction between error and uncertainty is that the error is a *point* estimate, while uncertainty is an *interval* estimate. Hence, error is different for each reading in a set, whereas uncertainty is same for the entire set of readings under given conditions. Although clause 2.2.4 of the GUM [1] clarifies that uncertainty can be viewed as 'a measure of possible error' in the estimated value of the measurand, here the error is to be interpreted as an admissible span and not as per its formal definition. Error is the cause of uncertainty. As error cannot be eliminated completely, the uncertainty always remains. As stated in clause 3.2.3 of the GUM, the terms 'error' and 'uncertainty' should be used properly and care should be taken to distinguish between them. The result of a measurement (after correction) may be very close to the value of the measurand and hence may have a negligible error but still it may have a large uncertainty. An *apparently* 'zero' error does not mean 'zero' uncertainty! Thus, the uncertainty of the result of a measurement should not be confused with the *remaining* part of the unknown error.

2. **Bias is not uncertainty**

Systematic error, often termed 'bias,' is a distinctly different concept. Sometimes correction is applied to a measurement result to compensate for a systematic effect. The uncertainty of that correction is *not* the systematic error, often termed bias. (For more details on error and bias, see chapter 5, 'Various error terms and bias.')

3. **Accuracy (or rather *inaccuracy*) is not uncertainty**

Accuracy is a *qualitative* term (expressed by adjectives like 'poor' or 'good') whereas uncertainty is a quantitative term. Uncertainty is the quantification of the quality of measurement. Accuracy is expressed with reference to a standard or reference value, while uncertainty is associated with the best estimate of the value of the measurand. (For more detail on accuracy, see chapter 1, 'Using correct terminology.')

4. **The standard error of the mean is not the uncertainty**

(Nor is it the uncertainty of random error.)

5. **A mistake is not the uncertainty**

A mistake is a slip-up, lapse, or gaffe, usually on the part of the operator. So the uncertainty is not a mistake in measurement. So a small (high) uncertainty does not mean small (big) mistake. However, mistakes in measurement process can result in errors, which will definitely affect the uncertainty.

6. **Product specifications are not uncertainty**

Product or equipment specifications are technical parameters and conditions under which measurement (in)accuracy is valid.

7. **Tolerances are not uncertainty**

Tolerances are the limits used to accept (or reject) a measurement result, including conformity decisions. Sometimes, in the specifications, the uncertainty may be expressed as the half-width of an interval that has a level of confidence stated with a $\pm$ sign. Thus, this uncertainty should not be perceived as tolerance.

8. **Uncertainty is not a way of assigning a safety limit or allowance to a quantity**

When a correction b is not applied to the result for known systematic effects for whatever reasons, this correction should not be used or accommodated so as to 'enlarge' the uncertainty. (Although the uncertainty needs to be factored in for conformity decisions as per ISO/IEC17025:2017 [2], there it is not envisaged from a 'safety' point of view.)

Clause F.2.4.5 of the GUM deals with a scenario in which a correction cannot be applied.

9. **Uncertainty is not same as 'expanded' uncertainty!**

Often, these two terms are used interchangeably by sheer misperception due to semantic similarity. The two terms have very distinct definitions and also differ in their values and character.

We are familiar with the definition of uncertainty given in the GUM (B.2.18) that defines it as a 'parameter, associated with the result of a measurement, that characterizes the dispersion of the values that could reasonably be attributed to the measurand.' On the other hand, the expanded uncertainty, as per the GUM (clause 2.3.5) is a 'quantity defining an interval about the result of a measurement that may be expected to encompass a large fraction of the distribution of values that could reasonably be attributed to the measurand.' To evaluate 'expanded' uncertainty, we need to estimate the 'uncertainties' of different contributions and combine them to get 'combined uncertainty.' This combined uncertainty is then multiplied by a 'coverage factor' (k) to get the expanded uncertainty.

Therefore, swapping these two terms should be strictly avoided.

10. **Uncertainty is not the error that remains after correction**

The GUM recommends that corrections should be applied to the measurement results for the known systematic effects to the extent possible, which may result in a very small error. This remaining small error after correction should not be confused with uncertainty. In spite of a negligible error is obtained for the quantity value, the uncertainty of that measurement could be very large!

11. **Finally, uncertainty is not a way to account for mistakes**

Mistakes, known errors or unknown errors, and blunders are inadvertent parts of human life. 'Mistake' is a general term which describes something that may happen while observing or recording data. It usually takes place at the subconscious level.

The term 'error' generally indicates the result of an incorrect application of mind, procedure, or method or deficient knowledge and usually happens at the conscious level.

A 'blunder' is the result of carelessness, negligence, or ignorance and is typically part of a person's gross traits.

Thus, uncertainty is none of the above and shall not be understood as such.

References and further reading

[1] JCGM 100:2008 *(GUM) Evaluation of Measurement Data—Guide to the Expression of Uncertainty in Measurement* 1st edn (BIPM, IEC, IFCC, ISO, IUPAC, IUPAP, OIML—International Organization for Standardization)

[2] ISO/IEC 17025:2017 *General Requirements for the Competence of Testing and Calibration Laboratories* (Geneva: ISO)

IOP Publishing

A Practical Handbook on Measurement Uncertainty
FAQs and fundamentals for metrologists
Swanand Rishi

Chapter 5

Various error terms and bias

Abstract: It is important to understand various error terms, since error is the cause of uncertainty. Proper management and control of errors are important to achieve lower uncertainty.

> *Everybody believes in the exponential law of errors:* **(i.e. the normal distribution)** *the experimenters, because they think it can be proved by mathematics; and the mathematicians, because they believe it has been established by observation.*
>
> —Gabriel Lippmann

Many types of error terms are used in metrology. The most common among them, error and bias, create some confusion and are used interchangeably in error. The other terms are systematic error and random error.

Error, as defined in the VIM (2.26) [1], is a 'measured quantity value minus a reference quantity value' and in the GUM (B.2.19) [2], it is the 'result of a measurement minus a true value of the measurand.'

Error is the arithmetic difference between *each* measured value and the true value of the measurand; the true value in practical scenarios being taken as the reference or standard value. (As clarified by the GUM in its note to the definition, since the true value cannot be determined, in practice a conventional true value is used.) Thus, if we take five readings, there are five errors because each reading is compared with the reference value. In contrast to the five errors, there is only *one* bias, namely the mean of the five readings minus the reference value.

As no measurement system (including the metrologist) is perfect, the imperfections give rise to errors which are traditionally viewed as having two components, namely: a *random* component of error and a *systematic* component of error.

Random error as per the GUM (B.2.21) is the 'result of a measurement minus the mean that would result from an infinite number of measurements of the same

doi:10.1088/978-0-7503-6462-1ch5

measurand carried out under repeatability conditions.' Thus, random error is equal to error minus systematic error. (Note that unlike the definition of systematic error, the definition of random error does not refer to a 'true' value.)

The VIM (2.19) defines random error as a 'component of measurement error that in replicate measurements varies in an unpredictable manner.'

The random error cannot be compensated or eliminated but can only be reduced by taking a large number (n) of readings. Random errors exhibit a Normal distribution for large n. The larger the sample size, the closer the distribution becomes to the normal. (If known influence factors are well controlled, even a small n can result in a Normal distribution.) The mean of a large number of readings is more realistic because a large set is supposed to include the effects of maximum influence factors, including temporal factors such as drift and stability. Further, for larger sample sizes n, the experimental standard deviation of the mean that represents the uncertainty also reduces, as it is inversely proportional to the square root of n.

Systematic error as per the GUM (B.2.22) is the 'mean that would result from an infinite number of measurements of the same measurand carried out under repeatability conditions minus a true value of the measurand.' (The term 'true' is redundant but used in the GUM for convenience and to aid in understanding the concept. To calculate systematic error, a conventional true value is used.) Thus, systematic error is equal to error minus random error. Systematic error is the component of measurement error that remains constant or varies in a predictable manner.

The VIM (2.17) defines systematic error as the 'component of measurement error that in replicate measurements remains constant or varies in a predictable manner.'

Thus, it equals measurement error minus random measurement error. It arises due to random and systematic causes and can be evaluated by experiments if the causes are known. Corrections can be applied for the known systematic errors.

The GUM (3.2.4) recommends that the results of measurements should be corrected for all known systematic effects. A correction (GUM B.2.23) is 'a value added algebraically to the uncorrected result of a measurement to compensate for systematic error. If they are not corrected, their uncertainty contribution shall be accounted for.' Of course, since systematic error cannot be known exactly, compensation for systematic error cannot be complete. It is assumed that after correction, the expected value of error arising from a particular systematic effect is zero. Of course, an unrecognized systematic effect cannot be taken into account in the evaluation of the uncertainty but nonetheless contributes to its error and also the uncertainty.

Figure 5.1 gives a pictorial presentation of random error and systematic error.

Bias is defined in the VIM (2.18) as an 'estimate of a systematic measurement error.' ISO 3534–1 defines it as 'the difference between the expectation of the test results from a particular laboratory and an accepted reference value.' It is often used as a synonym for systematic error. According to the VIM definition, bias has something to do with error. Bias is also called 'total systematic error,' implying that one or more systematic error components may contribute to it. It may be noted that when a correction is applied to a measurement result to compensate for a systematic effect, the uncertainty of that correction is *not* the systematic error, often called bias.

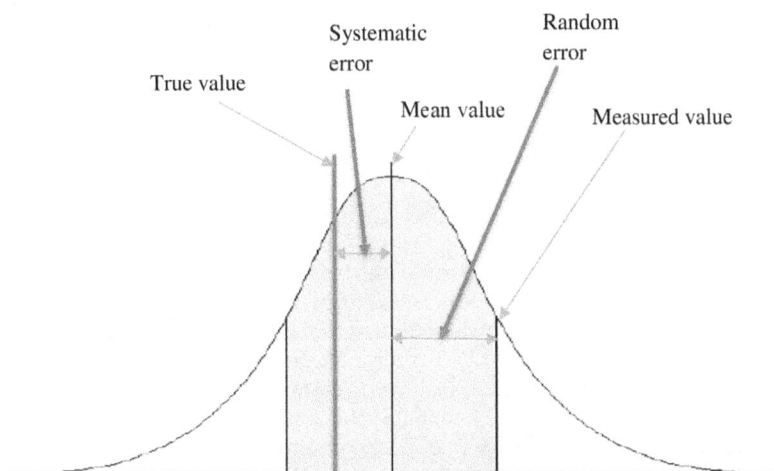

Figure 5.1. Random error and systematic error.

(**'Instrumental bias'** is separately defined in the VIM (4.20) as the 'average of replicate indications minus a reference quantity value.')

However, as the VIM defines systematic error quite elaborately with notes but defines bias only in connection with a measuring instrument, it is recommended to use the terms 'bias' and 'systematic error' distinctly.

The following example better illustrates the above concepts.

A set of five repeat measurements are taken using a DC voltmeter that measures a nominal output of 10 V produced by a calibrator as 10.001, 9.9996, 10.000, 9.9995, and 10.003 V. Let the assigned value at 10 V from the certificate of the calibrator be 10.0005 V with negligible uncertainty.

Thus, the mean value of the five measurements is 10.001 V (rounded to indicator resolution).

A more rigorous set of 30 readings gives a mean of 10.002 V. These measurements are carried out under controlled conditions; hence, we can assume that they correspond to infinite measurements.

Thus, there are five (random) errors (each reading of the voltmeter minus the mean of **30** readings) viz, -0.001 V, -0.00024 V, -0.002 V, $-0.000\,25$ V, and 0.001 V.

The 'bias' is (the mean of **30** readings minus the 'true' value) $= 10.002$ V-10.0005 V $= 0.0015$ V.

However, in almost all practical cases, it is not possible to take the mean of a large (equivalent to infinite) number of readings due to constraints on resources. Hence, the mean of the actual sample size is assumed to represent the mean of an infinite number of readings. In that case, the 'bias' is (the mean of **five** readings minus the 'true' value $= 10.001$ V-10.0005 V $= 0.0005$ V. In this example, the difference between the 'ideal' bias and the 'actual' bias is $= 0.0015-0.0005 = 0.001$ V.

The GUM also does not use terms 'Type A error' and 'Type B error,' which at times are wrongly used by some metrologists. Remember that 'Type A' and 'Type B' are *methods* of evaluating uncertainty.

A fact one should know

Error and uncertainty are distinctly different concepts and should not be swapped. Error is a qualitative term and an idealized concept, as it is calculated with reference to a standard (or conventional true) value that is not known exactly and is perennially unknowable. Thus, even if error or bias is zero, uncertainty is always present and has a finite nonzero value, however small. Prior to the new SI system of units adopted with effect from 20 May 2019, only a few fundamental constants of nature (e.g. the speed of light in vacuum, the permeability of free space) were supposed to be known *exactly* and hence had no uncertainty. Thus, with the old SI system of units, if one were conducting an experiment to measure the speed of light in vacuum, the error of measurement could be said to be *'known'* but would have some uncertainty.

Points to ponder

With reference to the SI units mentioned in the box above, it is worth digressing to a significant aspect of the new SI units adopted on 20 May 2019. As per the new definitions adopted, all units have been defined in terms of various constants of nature. The definitions of the base units specify the exact numerical value of each constant when its value is expressed in the corresponding SI unit. While four of the seven base units (the kilogram, ampere, kelvin, and mole) have revised definitions in the new SI, the definitions of the second, meter, and candela remain unchanged.

All units of the SI can be written either using a defining constant itself, or using products or quotients of the defining constants. Definitions based on defining constants are called *explicit-constant definitions*; which are based on the fundamental constants of nature. In contrast, *explicit-unit definitions* are based on particular experimental procedures.

The numerical values of these seven constants when expressed in SI units provide the definition of all SI units, both base and derived units.

As emphasized elsewhere, ISO/IEC 17025 requires that the measurement results reported by laboratories shall be traceable to SI units, unless demonstrated that it is not technically possible.

For the sake of knowledge, the numerical values of constants and the units they define are given below (adapted from the BIPM website).

*It may be noted that the numerical values of the seven defining constants have **no** **uncertainty!***

Defining constant	Numerical value	Unit
The cesium hyperfine frequency $\Delta\nu_{Cs}$	9 192 631 770	Hz
The speed of light in vacuum C	299 792 458	m s^{-1}
The Planck constant h	6.626 070 15 $\times 10^{-34}$	J s
The elementary charge E	1.602 176 634 $\times 10^{-19}$	C
The Boltzmann constant K	1.380 649 $\times 10^{-23}$	J K^{-1}
The Avogadro constant N_A	6.022 140 76 $\times 10^{23}$	mol^{-1}
The luminous efficacy of a defined visible radiation K_{cd}	683	lm W^{-1}

Readers are encouraged to visit BIPM's site https://www.bipm.org for more information, which metrologists will certainly find enticing. Metrologists in all fields must be familiar with the proper use of SI units, which are often used with a casual approach and total disregard for a meticulously designed system.

A fact one should know

It should be noted that the GUM has nowhere used the terms 'random uncertainty' and 'systematic uncertainty'. However, it does classify errors as 'random error' and 'systematic error,' which we have discussed above. As uncertainty is the result of evaluating error components, it cannot in itself be 'random' or 'systematic'! Actually, as mentioned in cl. E.3.6(c) of the GUM, 'it is unnecessary to classify components as 'random' or 'systematic,' as all components are treated in the same way' (i.e. treated as standard deviations).

The following paragraph from the GUM cl. E.3.6 is worth remembering:

Benefit c) (it is unnecessary to classify components as 'random' or 'systematic') is highly advantageous because such categorization is frequently a source of confusion; an uncertainty component is not either 'random' or 'systematic.' Its nature is conditioned by the use made of the corresponding quantity, or more formally, by the context in which the quantity appears in the mathematical model that describes the measurement. Thus, when its corresponding quantity is used in a different context, a 'random' component may become a 'systematic' component, and vice versa. (Italics in brackets added by the author for clarity.)

References

[1] JGCM 200:2012 *International Vocabulary of Metrology—Basic and General Concepts and Associated Terms* 3rd edn (Sèvres: BIPM)
[2] JCGM 100:2008 *(GUM) Evaluation of Measurement Data—Guide to the Expression of Uncertainty in Measurement* 1st edn (BIPM, IEC, IFCC, ISO, IUPAC, IUPAP, OIML—International Organization for Standardization)

IOP Publishing

A Practical Handbook on Measurement Uncertainty
FAQs and fundamentals for metrologists
Swanand Rishi

Chapter 6

The Guide to the expression of uncertainty in measurement (GUM)—the new approach

Abstract: The process of evaluating uncertainty in measurement has undergone a paradigm shift since 1995. The prior approach was based on the treatment of errors (random and systematic), while the GUM's [1] new approach is based on treating uncertainties themselves by the Type A and Type B methods of evaluation. It has radically shifted the focus from errors and true values to uncertainty evaluation based on frequency-based point of view of probability as well as 'degree of belief' approach of probability that an event will occur. However, the limitations of this approach also need attention.

> *Facts are stubborn things, but statistics are pliable.*
>
> —Mark Twain

Although metrologists are well versed in the evaluation of uncertainty, it is commonly observed that some metrologists equate the Type A and Type B methods of evaluation with the earlier methods of evaluation, viz, the random and systematic methods respectively. This is a gross mistake and needs to be understood in light of the radically different approach of the GUM (also referred as the *Guide*).

The bases of the *Guide* are Recommendation 1 (CI-1981) of the Comité International des Poids et Mesures (CIPM) and Recommendation INC-1 (1980) of the Working Group on the Statement of Uncertainties. It was first issued in 1995 by adopting an 'uncertainty approach,' instead of the 'error approach' (also called the traditional approach or the true value approach) being followed until then.

The 'error approach' was based on the objective of attaining a value as close as possible to a single definite value or the so-called 'true' value. The errors were classified as random (or imprecise) and systematic errors (or bias or accuracy) and were treated and presented independently, since rules for combining them were not standardized. The terms 'error' and 'uncertainty' were used interchangeably due to

doi:10.1088/978-0-7503-6462-1ch6

the lack of a definition for uncertainty. The terms 'bias' and 'systematic error' were treated as synonymous. Further, a true value was considered 'unique' in principle, but practically unknowable. The idea of believing in a perfectly known value was like playing God. Due to the lack of harmonization, every laboratory used to estimate uncertainty in a different way, leading to difficulty in comparing results.

Post 1970, a need for the harmonization of uncertainty evaluation was recognized due to:

- increased international comparisons and conformity requirements;
- trade requirements;
- technological advances in almost all areas, particularly in electronics;
- increased awareness of the importance of uncertainty among leading calibration laboratories and National Metrology Institutes (NMIs).

It was thus felt that a uniform method was required which would be universal for all types of measurements and consider all input quantities. The quantity that expressed uncertainty was to be internally consistent; meaning it was to be:

- directly derivable from its constituent sources;
- independent of the type, grouping, or decomposition of components;
- transferable from one result to another;
- able to provide a realistic interval for the results of measurement, with some coverage probability that would encompass a large fraction of the distribution of values usually observed in industrial and commercial activities.

Recommendation INC-1 (CI-1980) of the Working Group on the Statement of Uncertainty and Recommendation 1 (CI-1981) of the CIPM was the basis for the guidelines that met all the above criteria. It was approved by the CIPM in 1981 and reaffirmed in 1986. Finally, the ISO, along with six other concerned bodies, published the *Guide,* more fully titled 'Evaluation of measurement data—Guide to the expression of uncertainty in measurement,' or the GUM for short. This produced a tectonic shift in the outlook towards the evaluation process of uncertainty from an 'error approach' to an 'uncertainty approach.' It is also called the GUM uncertainty framework (GUF).

The 'uncertainty approach' acknowledges that owing to the inherently incomplete amount of detail in the definition of a quantity, there is no single true value but a set of values consistent with the definition. However, this set of values is, in principle and in practice, unknowable. Only in special case of fundamental constants is the quantity considered to have a single true value (e.g. the speed of light or the permeability of free space—see the box 'Points to ponder' in chapter 5, 'Various error terms and bias' on the revised outlook of the SI units). The approach focuses on mathematical modeling of the measurement process together with the use of the statistical concept of probability with its recognized interpretations—both as 'frequency based' as well as 'degree of belief.' The GUM very aptly calls it an 'operational' approach intended to remove the often confusing relation between 'uncertainty' and the unknowable quantities 'true value' and 'error.' As it focuses on observed (or estimated) values of quantity and their dispersion, any mention of error becomes entirely unnecessary. It is based on the solid foundation of input and output quantities known as probability density functions or PDFs. The GUM approach is called a 'bottom-up' approach due to its emphasis on individual input quantities as a starting point for uncertainty evaluation.

It does, however, acknowledge that the model could be incomplete, as it is based upon available knowledge only, and it should be revised when found inadequate. The approach is to assign a reasonable interval, however small, to the measurand and acknowledge that the interval cannot be reduced, in principle and practice, to a single definitive value. This approach takes cognizance of the 'definitional uncertainty' associated with the measurand but definitional uncertainty is considered to be negligible compared to the other components of the uncertainty. Hence, the definitional uncertainty places a minimum limit on the uncertainty interval that is assigned to the result. If it is considered to be negligible, then the measurand is deemed to have an 'essentially unique' true value, also called the 'conventional true value.' Thus, in the GUM, the word 'true' is considered to be redundant and is retained only for the purpose of explanation of the concept and because of its prevalent usage.

Table 6.1. A comparison of the pre-GUM approach and the GUM approach.

Feature	Pre-GUM approach	GUM approach
Target value	True value	Best estimate of quantity value
Focus	Attaining closest true value	Interval around best value
Approach	Error based	Uncertainty based (also called a modeling approach)
Concept of target value	Single definite value in principle, but unknowable in practice	A range of values in principle, and unknowable in practice
Treatment of components	Independent or separate; cannot be combined.	Not independent or separate but identical; can be combined.
	Classified by nature of effects as 'random' or 'systematic'	Classified into 'Type A' and 'Type B' by method of evaluation
Basis for combination	Not applicable	Law of propagation of uncertainty
Method for combining uncertainties	Either linear addition, U_{add}, or RSS (root sum square) method, U_{rss}	RSS method
Degrees of freedom	Not applicable or realized	Provides formulae for degrees of freedom for Type B factors as well as effective degrees of freedom for combined uncertainty
Further/future incorporation of uncertainty	Not possible	Possible in a simple and logical way

(*Continued*)

Table 6.1. (*Continued*)

Feature	Pre-GUM approach	GUM approach
Other aspects	• Difficult to compare uncertainty with results of other measurements • Not based on mathematical modeling • Emphasis on experiment (relative frequency-based probability) • No concept of definitional uncertainty, hence no lower limit for uncertainty • No concept of expanded uncertainty	• Easy comparison of uncertainty with results of other measurements • Based on mathematical modeling (hence called a modeling approach) • Emphasis on experiment as well as knowledge (probability as 'degree of belief') • Definitional uncertainty acknowledged, which gives a lower limit for uncertainty • Concept of expanded uncertainty is introduced

Table 6.1 shows a feature-wise comparison of the two approaches.

Advantages of the new approach:
By foregoing the old model and adopting the new approach based on *method*, the GUM marked a paradigm shift in the process of evaluating uncertainty. The major advantages of the method are:
- It is possible to utilize the combined standard uncertainty of a result to evaluate another combined standard uncertainty in which the first result is used.
- The combined uncertainty can serve as a basis for intervals corresponding to realistic confidence levels
- All components can be treated identically, and there is no need for classification based on their so-called nature 'random' and 'systematic'. Classification is based on the context in which components appear in the mathematical model and hence may be swapped, depending on context. *This underscores the fact that no component by nature is either 'random' or 'systematic'.*

A fact one should know

Although the Type A and Type B methods of evaluation are recommended in the GUM, strictly speaking, the GUM (clause E.3.7) grants that there is no need to classify components as far as the calculation of combined standard uncertainty is concerned. The classification by method is not meant to distinguish components by their nature, i.e. random or systematic. However, it recognizes that such labels are useful for conceptualization and hence does provide a scheme for classifying methods. Nonetheless, classifying *methods* rather than *components* does not preclude us from gathering or grouping them in specific groups.

Although pre-GUM approach and GUM approach provide the same result of uncertainty, the GUM approach (Type A and Type B methods of evaluation) distinguishes the way we can *view* the components.

Also see the last 'Facts one should know' box in chapter 5 'Various error terms and bias.'

Limitations of the new approach (GUM uncertainty framework i.e. the GUF):
Clause 3.18 of JCGM 101 defines the GUF as the 'application of the law of propagation of uncertainty and the characterization of the output quantity by a Gaussian distribution or a scaled and shifted *t-distribution* in order to provide a coverage interval.'

Although the new approach overcomes many limitations of the old approach, it is nonetheless not devoid of limitations. (Accepting limitations is a hallmark of scientific inquiry or temperament. Again, nothing is perfect!)

The following are the limitations of the new approach

A) The GUF is valid for linear functional models if:

- at least one uncertainty contribution has finite degrees of freedom (so that the Welch–Satterthwaite formula can be applied);
- all input quantities are independent if they have finite degrees of freedom;
- the PDF of output Y can be approximated by a Gaussian, scaled, and shifted *t*-distribution (i.e. when the conditions of the central limit theorem are satisfied).

The above conditions are usually satisfied by routine measurements made in tier II/III and working-level laboratories.

B) The GUF is valid for nonlinear functional models if the conditions in clause 5.8 of JCGM 101 [2] are satisfied. These conditions are not usually encountered in routine measurements. They are complex in nature, difficult to evaluate, and hence not discussed in detail here.

References

[1] JCGM 100:2008 *(GUM) Evaluation of Measurement Data—Guide to the Expression of Uncertainty in Measurement* 1st edn (BIPM, IEC, IFCC, ISO, IUPAC, IUPAP, OIML—International Organization for Standardization)

[2] JCGM 101:2008 *Evaluation of Measurement Data—Supplement 1 to the Guide to the Expression of Uncertainty in Measurement—Propagation of Distributions Using a Monte Carlo Method* 1st edn (Sèvres: BIPM)

IOP Publishing

A Practical Handbook on Measurement Uncertainty
FAQs and fundamentals for metrologists
Swanand Rishi

Chapter 7

The law of propagation of uncertainty

Abstract: The entire edifice of the GUM [1] approach is founded on the law of propagation of uncertainty (LPU). It treats all components of uncertainty in terms of the standard deviations of their respective distributions. Although quite academic, it is worth understanding the law and the conditions under which it holds true.

As far as the laws of mathematics refer to reality, they are not certain; and as far as they are certain, they do not refer to reality.

—Albert Einstein

The central limit theorem (CLT) helps us convolve different distributions under certain assumptions. (See chapter 14, 'How is it that we can combine different distributions?') A further law that allows us to combine input uncertainties to get the uncertainty of the output is the 'law of propagation of uncertainty.'

The combined standard uncertainty (for uncorrelated input quantities, which is a common case) is the positive square root of the combined variance $u_c^2(y)$, which is given by

$$u_c(y) = \sqrt{\sum_{i=1}^{N}\left(\left[\left(\frac{\partial f}{\partial x_i}\right)^2 u^2 x_i\right]\right)} \tag{7.1}$$

where

f is the function given in $Y = f(X_1, X_2, ..., X_N)$;

Y is the measurand determined from N other quantities $X_1, X_2, ..., X_N$ using the functional relationship f given above;

$\frac{\partial f}{\partial x_i}$ is the *partial* derivative w.r.t. x, and

$u^2 x_i$ is the variance of x_i.

doi:10.1088/978-0-7503-6462-1ch7

This equation is called the LPU and is based on a first-order Taylor series (i.e. linear) approximation of the measurand Y. (Equation (7.1) is often called the 'general law of error propagation.')

When the nonlinearity of the function $Y = f(X)$ (where X stands for the input quantities) relating input and output quantities is significant, higher-order terms must be included in the equation.

(The above equation is for uncorrelated input quantities—a common case in routine measurements. Additional terms are included for correlated input quantities.) Thus, the combined standard uncertainty is the *linear* sum of the standard uncertainties of the input estimates. The above equation (equation 10 in clause 5.1.2 of the GUM) also emphasizes that all input quantities are treated as having same nature and are considered identically, irrespective of their origin or method of evaluation.

The statement of the LPU: 'The uncertainties of the input quantities, taken equal to the standard deviations of their probability distributions, combine to give the uncertainty of the output quantity, if that (output) uncertainty is taken equal to standard deviation of its probability distribution.'

In simplified form, the above statement is modeled mathematically as

$$u_c^2(y) = \sum_{i=1}^{N}[c_i u(x_i)]^2, \tag{7.2}$$

where c_i is the sensitivity coefficient.

This implies that the law applies to the propagation of multiples of *coverage factors* (being constant), but not for confidence intervals. (In deriving this equation, it is not assumed that the quantities have a Normal distribution). In this sense, the ISO Guide 98-3 (GUM) [1] propagates uncertainties that are evaluated from *estimates* based on assumed probability distributions rather than the distributions themselves. (Investigations based on the distributions of each influence quantity, instead of *estimates* of those distributions, are not economical and thus not worth doing). The ISO Guide very categorically states that the interval $[y \pm ku(y)]$ is not to be interpreted as a confidence interval and that the coverage probability (level of confidence) is not the confidence level of the *classical* statistics. The *Guide*'s definition of coverage probability is rather aligned with *Bayesian* statistics. (This is one of the contentions on which statisticians have contrasting views, but we need not digress for that here. It is enough to note that the confidence interval *does not indicate the probability* that the observed interval contains the true value of the parameter: it either has it or does not have it.)

Thus, the LPU equation requires the input quantities to have a Normal distribution and requires the function to be linear; if it is not linear, the higher-order terms must be negligible. As elaborated in the note to clause E.3.3 of the GUM, in earlier evaluations of uncertainty, this was perhaps the reason for separating (and not combining) the 'random' components (based on observations supposed to have a Normal distribution, i.e. a classical view of probability) and 'systematic' components (supposed to have non-Normal distributions with upper and lower bounds, i.e. a nonclassical probability view). However, this hesitation is overcome using the CLT, albeit under some conditions, as discussed in chapter 14 'How is it that we can combine different distributions?'

An important advantage of the LPU is that adopting it makes it possible to employ the uncertainty of one result in another result to calculate the combined uncertainty of that result.

Facts one should know

FACT 1: The advantage of the LPU is that it is possible to utilize the combined standard uncertainty of a result in the evaluation of another combined standard uncertainty in which the first result is utilized. For example, let the relative uncertainty of a reference standard of laboratory ABC calibrated by a higher laboratory be 0.05%. If laboratory ABC uses this standard for the measurement of its customer's device, this uncertainty can be used as one component (Type B) of uncertainty. If 0.05% uncertainty is given at a 95% confidence level and the coverage factor $k = 1.96$, its contribution to the current measurement is $0.05\%/1.96 = 0.026\%$. Also, in comparison measurements or where a single reading has to be taken, we can utilize the pooled standard deviation as the uncertainty of the process or method, provided the method is similar.

FACT 2: The statement of the LPU underscores the significance of standard deviation as a very dominant statistic in uncertainty evaluation. Both input and the output estimates have to be obtained in terms of standard deviation to signify the variability of data—both observed as well as heuristic. See chapter 8, 'Why is standard deviation used instead of variance?' for further elaboration.

Reference

[1] JCGM 100:2008 *(GUM) Evaluation of Measurement Data—Guide to the Expression of Uncertainty in Measurement* 1st edn (BIPM, IEC, IFCC, ISO, IUPAC, IUPAP, OIML—International Organization for Standardization)

Chapter 8

Why is standard deviation used instead of variance?

Abstract: Standard deviation is one of the most popular and common statistics used to measure variation, although it is not the basic term in statistics. The basic term used as a measure of variability is 'variance.' The two terms are, however, mathematically related.

> *Statistics may be defined as 'a body of methods for making wise decisions in the face of uncertainty.*
>
> —W A Wallis

'Standard deviation' is perhaps one of the most frequently used terms in statistics, apart from 'mean.'

This subject can be dealt with in two parts:

- Why do we consider standard deviation instead of range, which is also a measure of dispersion?
- Why is standard deviation used instead of variance, even though the latter is the more fundamental term in statistics?

The first point is more elementary and statistical in nature. All metrologists are familiar with standard deviation as a measure of the dispersion (or variation or spread) of measurement results. There are other measures of variation, such as range, quartile deviation, and mean deviation. Among these, the 'range' is the simplest to calculate. However, it is not preferred in uncertainty estimation (and many other statistical analysis) because:

1. By definition, the range = the maximum value − the minimum value. Thus, it takes into account only the minimum and maximum values in a data set. Hence, not all data points in the set get represented. Further, as only the

extreme points are taken, the distribution or composition of points *within* the range is of no consequence.

2. Either or both of these two values (maximum and minimum) may themselves be outliers (data points in a set that are far away from the general spread) and may make the range practically meaningless. (See chapter 31, 'Sample analysis—detecting the outliers' in Part F.)
3. It usually shows wide fluctuations from sample to sample.
4. It is not amenable to mathematical treatment.

Quartile deviation and mean deviation find little applications in statistics of measurement uncertainty. Hence, we shall move to the most important and widely used statistic of variability, which is 'standard deviation.'

The 'experimental standard deviation of the **sample data'** is given by

$$s(x_i) = \sqrt{\left\{\sum_{i=1}^{n}(x_i - \bar{x})^2/(n - 1)\right\}} \tag{8.1}$$

where n is number of data points or samples, x_i is the ith reading, and $\bar{x}$ is the mean of n readings.

The standard deviation of the **mean** of the sample data is given by

$$s(\bar{x}) = \frac{s(x_i)}{\sqrt{n}}. \tag{8.2}$$

$s(\bar{x})$ is also called the 'experimental standard deviation of the mean' (ESDM).

Note: the ESDM should not be called the 'standard error of the mean,' even though the standard error implicitly refers to the standard deviation of the sample mean.

The standard deviation (of the mean) is the most powerful measure of dispersion because:

1. It takes into account *all* data points.
2. It is least affected by fluctuations of samples. (Outliers make very little impact on its value compared to their impact on range.)
3. It is amenable to mathematical treatment.

The only drawback of standard deviation is that it gives more weight to extreme points.

The term under the square root in equation (8.1) is called 'variance.'

The GUM [1] (at C.3.2) defines the variance (of a random variable) as 'the expectation of its quadratic deviation about its expectation,' and as per C.2.11 in JCGM 100:2008 (clause 1.22 of ISO 3534-1:1993), it is the 'expectation of the square of the centred random variable.' In statistical parlance, variance is the 'moment of order r, where r equals 2 in the centred probability distribution of the random variable.'

Therefore, variance is a more fundamental term in statistics that represents dispersion or variation, and the standard deviation is *derived* from variance simply by taking its positive square root. (Standard deviation is defined as such in the GUM section C.2.12. Note that variance is first defined in the GUM at C.2.11 and then standard deviation is defined at C.2.12. The same is also observed in ISO 3534-1 [2]. It may further be noticed that variance is not defined as the square of the standard deviation!) Variance and standard deviation are further defined in the GUM at C.2.20 and C.2.21, respectively. The definition of variance at C.2.20 is practical definition that lends itself to its formula.

Fact one should know

To understand the second point, let us see what variance means. Note that variance is a specific term *defined* in statistics, and standard deviation is defined as 'a positive square root of variance.' *So, variance is the fundamental term in statistics.*
  Mathematically, the variance is given by

$$\sum_{i=1}^{n}(x_i - \bar{x})^2/n - 1$$

(This is the same term as in equation (8.1), but without 'square root.') Thus, in concurrence with this equation, the GUM in clause C.2.20 defines it as 'a measure of dispersion, which is the sum of the squared deviations of observations from their average divided by one less than the number of observations.'
  If both variance and standard deviation are interrelated by just an exponent (square or square root), it should not actually matter whether we use either of the two! However, there is a catch. In quantitative metrology, we invariably have some unit of measurement for the measurand. If we use the above formula as a measure of dispersion, by dimensional analysis, the resultant unit will be the square of the basic unit. This will be difficult to comprehend and confusing in certain cases. For example, if we are measuring length, the base unit is the meter, but the unit of variance is the 'square meter' (m^2), which is the unit of area. If we express the mean length of a rod as 1.003 **m** with a variance of, say, 4×10^{-6} m^2, it is hard to understand this due to the use of different units. The correlation between length in meters and dispersion (variance) in square meters is difficult to grasp. Instead, in the above example, if dispersion (variance) is expressed in terms of standard deviation = 2×10^{-3} m = 0.002 m, one can easily correlate the length and its dispersion (variance), as both have the same unit.

Another important reason (for using standard deviation as a measure of dispersion) is that we are able to combine Type A and Type B uncertainties due to the *'law of propagation of uncertainty'*, which utilizes concepts of 'variance' and 'standard deviation.' Obviously, no other measure of dispersion (e.g. range) is pertinent in the evaluation of uncertainty.

A fact one should know

FACT 1
All definitions of variance use the term 'mean' or 'expectation' (also called expected value). The mean is one of the measures of the central tendency, the others being the mode (the most frequently occurring value) and the median (the value of the middle

---

item when items are arranged in ascending or descending order). Among these, the mean is believed to be and empirically observed to be the best estimate of the expectation in most applications. Thus, variance and hence the standard deviation of the *mean* represent the spread or dispersion of the sample data.

When the sample size is large, the distributions of both the mean and the median are normal. However, it has been statistically proved that the standard deviation of the median is 1.25 times the standard deviation of the mean (which is given by $s/\sqrt{n}$; s being the sample standard deviation). Hence, the standard deviation of the mean is a better estimator of the population mean, since it is *likely* to be closer to it. Statistically, it has also been found that values of all other estimators of dispersion are higher than the standard deviation of the mean. Thus, the mean is the most versatile, robust, and efficient estimator of the central tendency.

FACT 2

The term 'mean' (i.e. the '*arithmetic*' mean by default) is generally used when referring to a **population** parameter, and the term 'average' is used when referring to the result of a calculation on the data obtained in a **sample** (refer to NOTE 1 at C.2.19 of the GUM).

The average of a simple random sample taken from a population is an unbiased estimator of the mean of this population. However, there are other estimators, such as the geometric or harmonic mean, in addition to the median or mode.

References

[1] JCGM 100:2008 *(GUM) Evaluation of Measurement Data—Guide to the Expression of Uncertainty in Measurement* 1st edn (BIPM, IEC, IFCC, ISO, IUPAC, IUPAP, OIML—International Organization for Standardization)

[2] ISO 3534-1:2006 (confirmed in 2021) 2006 *Statistics — Vocabulary and symbols — Part 1: General statistical terms and terms used in probability* (Geneva: ISO)

IOP Publishing

A Practical Handbook on Measurement Uncertainty
FAQs and fundamentals for metrologists
Swanand Rishi

Chapter 9

Using pooled standard deviation

Abstract: The pooled standard deviation is a type of data that is already available from earlier experiments and which can be effectively used in the current experiment. This reduces the burden of evaluating Type A uncertainty to some extent by simplifying the calculations of the current experiment that is carried out using the previous procedure.

> *In God we trust. All others must bring data.*
>
> —W Edwards Deming

We know that the standard deviation of the error component represents the uncertainty contribution due to that error, regardless of whether Type A or Type B evaluation is used. In most Type A evaluations, we take a series of measurements to evaluate the mean as well as the standard deviation of the mean. This is usually done when actual measurements are involved. However, there are situations in which comparison methods are used, e.g. temperature sensor calibration (except that using fixed-point cells), weight calibration, end gauge calibration. Another example is hardness testing, in which the strict repetition of an observation is impossible due to the different *physical* locations of the observations. (One has to take repeat readings at different spots for hardness measurement.) In such measurement scenarios, the unit under test (UUT) is compared with a reference standard using a common measurement system.

UKAS M3003:2007/2024 [1] clause 4.10/4.7 states that 'it may not always be practical or possible to repeat the measurement many times during a test or a calibration. In these cases, a more reliable estimate of the standard deviation of a measurement system may be obtained from data obtained previously, based on a larger number of readings.' UKAS M3003:2007 also advises 'to take at least two measurements; however, it is acceptable for a single measurement to be made even though it is known that the system has imperfect repeatability, and to rely on a previous assessment of the repeatability of similar devices.'

In day-to-day measurements, instead of going through the rigor of evaluation of standard deviation, one can make use of the standard deviation already obtained from a well-established measurement system. The standard deviation estimated from such a system is based on a comparatively large sample size, typically 10–20 readings. This standard deviation, called 'pooled standard deviation,' s_p, which is estimated using a well-characterized system with a larger sample size, is a better approximation for use in other similar measurements carried out later. The standard uncertainty in this case is given by

$$u = s_p / \sqrt{n},$$

where n is the sample size of the *current* measurement.

An example from balance calibration will illustrate this concept.

The calibration of a weight of 10 kg of OIML class M1 is done by comparison to a reference standard using a mass comparator with known performance characteristics (i.e. pooled standard deviation). The measurement sequence follows the A-B-B-A method, in which A and B are the standard weight and the weight under calibration, respectively.

The observed readings are given below. (The readings after the decimal point are only shown for convenience.)

No.	A reading (g)	B reading (g)	B reading (g)	A reading (g)	Observed difference (g)
1.	0.015	0.020	0.025	0.010	0.010
2.	0.010	0.030	0.020	0.010	0.015
3.	0.020	0.045	0.040	0.015	0.025
4.	0.020	0.040	0.030	0.010	0.020
5.	0.010	0.030	0.020	0.010	0.015

Assuming that the well-characterized method used in the above experiment is known to give a pooled standard deviation of comparator repeatability of 30 mg, the Type A uncertainty is

$$U_A = 30/\sqrt{n}$$
$$= 30/\sqrt{5} = 13.42 \text{ mg},$$

where n is the number of sets of repeat measurements.

The point to be noted is that we do not have to separately calculate Type A standard uncertainty for the observed readings. This saves time and effort.

Thus, in such cases, there is no need to estimate the standard deviation of the current data, since the pooled standard deviation s_p can be used instead. This technique could be best put to use for routine measurements and cases in which the TUR is much higher than 4:1, thereby reducing the cost and time of laboratory operations. However, care should be taken to ensure that the measurement process followed is same as that used to estimate the pooled standard deviation in the past. It is also essential to periodically review the past process and confirm the validity of s_p.

A previous estimate of standard deviation can only be used if there is no further change in the measurement system/procedure that could have an effect on the repeatability of the current measurement. If there is any doubt about the measured values (raised by a deviation from the typical values of the measurement system), the causes should be investigated and resolved before proceeding further.

A fact one should know

In a more rigorous characterization process, a pooled estimate of variance s_p^2 based on N series of independent observations of a random variable is given by:

$$s_p^2 = \frac{\sum_{i=1}^{N} \nu_i s_i^2}{\sum_{i=1}^{N} \nu_i},$$

where s_i^2 is the experimental variance of the ith series of n_i independent repeated observations and has $\nu_i = n_i - 1$ degrees of freedom. The degrees of freedom of s_p^2 are $\nu = \sum_{i=1}^{N} \nu_i$. The experimental standard deviation $s_p/\sqrt{m}$ of the arithmetic mean of m independent observations taken later, characterized by the pooled estimate of variance s_p^2, also has ν degrees of freedom. Clause 4.3.1 of the GUM calls this component of uncertainty 'previous measurement data,' under the Type B method of evaluation of uncertainty. However, it is also used for the evaluation of Type A uncertainty.

It should be noted that this approach is valid only if the procedure followed for the current observations is very similar to the well-characterized measurement system for which the pooled standard deviation is known.

Reference

[1] UKAS M3003:2007/2024 *The Expression of Uncertainty and Confidence in Measurement* (Staines-upon-Thames: United Kingdom Accreditation Service)

IOP Publishing

A Practical Handbook on Measurement Uncertainty
FAQs and fundamentals for metrologists
Swanand Rishi

Chapter 10

Some uncommon uncertainties

Abstract: Apart from common terms such as uncertainty, standard uncertainty, and expanded uncertainty, there are quite a few uncommon uncertainty terms which may have to be accounted for, where applicable and relevant.

> *The perfect is the enemy of the good.*
>
> —Voltaire

Generally, we encounter a few sources of uncertainty in Type B evaluation, perhaps four on average. In addition to repeatability (i.e. Type A evaluation), clause 3.3.2 of the GUM [1] lists nine possible factors under Type B evaluation, although not all of them necessarily contribute concurrently in routine measurements. The typical factors that contribute under Type B evaluation as listed under clause 4.3.1 of the GUM are:

1. Previous measurement data. This can be used in typical comparative measurements as a pooled standard deviation s_p. (See chapter 9, 'Using pooled standard deviation.')
2. Experience with, or general knowledge of, the behavior and properties of relevant materials and instruments. This helps in the assignment of a particular distribution to the data.
3. Manufacturer's specifications: the uncertainty and other factors such as linearity, drift, stability, temperature coefficient, etc. are available from the specifications.
4. Data provided in calibration and other certificates: the uncertainty, coverage factor at a given confidence level, and other data obtained from characterization (if any), are available.
5. Uncertainties assigned to reference data taken from handbooks: handbooks and research papers on specialized subjects give the uncertainties assigned to the reference data or results obtained.

From practical perspective, the above sources of uncertainty are usually sufficient to estimate uncertainty. However, some uncommon effects mentioned in the GUM also contribute to uncertainty, depending upon the application. Let us take a look at those factors:

1. **The uncertainty of finite-precision arithmetic:** In some measurement systems, computers are used for data logging and other arithmetic calculations. Processors and ADCs have limited word lengths, say 32 bit or 64 bit. The algorithms used have a much greater capability to process data, but it is truncated due to the use of finite resolution. If δ_x is the smallest change in the output quantity that can be empirically obtained by changing the most important input quantity, then δ_x can be taken as a contributory factor. Assuming it has a rectangular distribution, its uncertainty contribution is $0.29\delta_x$.

2. **The uncertainty of the method of measurement:** This is one of the most difficult components to establish. IEC's Electropedia 311-2 (for Electrical and Electronic measurements) provides nine different methods of measurement: direct, indirect, comparison, substitution, complementary, differential, null, beat, and resonance.

   Just as an example, the measurement of a one-ohm resistor gives different results when the measurement takes place using a 4–1/2 digit DMM, a 6–1/2 digit DMM, or a separate current source and a voltmeter (the I–V method). In cases other than that of the 4–1/2 digit DMM, a two-wire and four-wire method would give different results. It should be noted that the uncertainties in each case would also be different.

   Generally, measurement methods and procedures are well established and all known influence factors are controlled. However, the method of measurement does affect the result as well as the uncertainty. A different method may be conceived and subjected to trials, and if it shows significantly less variability, it may be used as a standard procedure after due validation. Such a new method can be treated as an *a priori* probability distribution and may be a dominant contributor to uncertainty. All these methods may have some inherent common factor, e.g. 'instrument offset,' that results in variability, and it can only be evaluated based on one's existing knowledge. An interlaboratory comparison with the freedom to use independent methods could be used to determine the contributions of the different methods.

   (See chapters 13 'Reproducibility in uncertainty evaluation' and 34 'Alternative approaches in uncertainty evaluation' for further elaboration of this topic.)

3. **The uncertainty of the sample:** In many measurements, an unknown device is usually a material measure or a physical artifact. It is compared with a standard which has the same or similar characteristics. Some natural materials and chemical products that are measured are inhomogeneous. The sampling error (which shows how well the sample represents the population) and the sample treatment (the preparation of the sample for

measurement or analysis) also contribute to the uncertainty. In chemical measurements of a characteristic, it is possible that some other unanalyzed characteristics of a particular method may affect the result. The treatment of such factors requires skill, proper reasoning, and knowledge on the part of the metrologists.

According to the standard ISO/IEC 17025:2017 [2], the uncertainty of the sample, sampling process, and specimen treatment all become important considerations, since this standard includes sampling laboratories within its scope in addition to the usual test and calibration laboratories. Readers working in the fields of analytical chemistry, microbiology, food technology, etc. may consult [3] for further detail on sampling and its uncertainty. Another document for general sampling is [4].

4. **Uncertainty due to secular drift:** This term is used as a measure of long-term and nonperiodic variation over time. Whether it is secular stability or not depends upon the information available in the given timescale. A variation that appears nonperiodic over some short period may turn out to be periodic over a sufficiently longer period. This is a characteristic particularly applicable to material measures or physical artifacts such as passive standards, e.g. a resistor or a weight. Starting with the manufacturer's specifications, the data (e.g. calibrated values from a certificate) can be analyzed over some period. A curve fitting would be required to find the rate of drift, which would be used as a component of uncertainty with an assumed rectangular distribution.

Stability is related to variations that take place over short periods and is typically used when the source or stimulus shows fluctuations. Section 4.19 of the JCGM 200:2012 (also known as the VIM [5]) defines the stability of a measuring instrument as a 'property of a measuring instrument, whereby its metrological properties remain constant in time.' However, it is also applicable for source equipment. The most common examples are heat sources such as liquid baths, dry blocks, or furnaces. The stability is determined by taking 8–10 relatively stable readings after stabilization.

5. **The uncertainty of interpretation:** Sometimes, interpolation has to be performed for intermediate values, since calibration is typically carried out at three to five points per range. In temperature measurements, this is very common and requires curve fitting. The nonlinearity of the curve, scale corrections, or deviations from the reference data introduce uncertainty due to possible different interpretations. This may also happen in the case of different results obtained by researchers for the same quantity under similar conditions.

6. **Definitional uncertainty:** This uncertainty arises due to our limitations in defining a quantity completely. It is also termed 'intrinsic uncertainty' in the GUM and IEC 60359 [6]. As per clause 2.27 of the VIM, it is 'component of uncertainty resulting from the finite amount of detail in the definition of a measurand.' Before any measurement is performed, the measurand—the quantity to be measured—has to be defined to the maximum possible extent.

However, in principle, a measurand cannot be *completely* described without an infinite amount of information.

The GUM gives an example in Annexure D.1.2: the velocity of sound in dry air of composition (mole fraction) $N_2 = 0.7808$, $O_2 = 0.2095$, $Ar = 0.009\ 35$, and $CO_2 = 0.000\ 35$ at the temperature $T = 273.15$ K and pressure $p = 101\ 325$ Pa.

However, in spite of giving so much description (or so many conditions) in the above definition, it is recognized that there could still be some unknown influence factors (e.g. humidity, dust) that could influence the measurand. Thus, a measurand cannot be *completely* described without an infinite amount of information, leaving scope for interpretation and thus introducing uncertainty of some small value. Hence, definitional uncertainty is the minimum uncertainty practically achievable under a given system of measurement. Obviously, its value changes as soon as the definition of the measurand changes. The GUM assumes that this uncertainty is negligible compared to other sources of uncertainty. If it is significantly large, it recommends that it should be included in the evaluation of uncertainty in measurement. (Estimating the value of definitional uncertainty is a very tough task, however. Fortunately, this is a concern for the BIPM and other apex bodies!) The 'true' value is the one that satisfies the definition completely, but since the definition itself has some uncertainty, the true value is indeterminate. Because of this definitional uncertainty alone (and a focus on the best estimate of the measured value), the concept of 'true' value is discarded in the GUM.

Note that any change (while improving the definition) in the descriptive details of the measurand leads to another definitional uncertainty. In clause D.3.4 of the GUM and in IEC 60359, the concept of 'definitional uncertainty' is known as 'intrinsic uncertainty.' However, as explained in chapter 5 'Various error terms and bias,' *the numerical values of the seven defining constants have **no uncertainty!***

7. **Uncertainty of correction:** Due to effect of external influence factors, the mean of the sample data may have some bias due to systematic effects. The VIM defines 'correction' as 'compensation for an estimated systematic effect.' The GUM assumes that such results are corrected for all known systematic effects (with the further assumption that efforts to identify them are taken in good faith!). However, even if correction reduces the bias to zero, the uncertainty of correction remains. This uncertainty of correction should *not* be called systematic error or bias.

A fact one should know

The GUM recommends that correction should be applied except when the cost of such activity is prohibitively high. It suggests a simpler approach in F.2.4.5, if certain conditions are fulfilled. However, this approach employs a single average correction that is applied to all readings in the range (e.g. in the case of characterization of a temperature sensor over a range) and may lead to overstatement or understatement of the individual results in the range.

Second, a practice followed by some metrologists of not applying correction '*b*' but accounting for it by enlarging the uncertainty by the same amount is deprecated in the GUM. This correction '*b*' is usually a rectangular distribution and should not be algebraically (or otherwise) added to the uncertainty, which is assumed to be a Normal distribution. Thus, the purpose of evaluating uncertainty should not be to assign a *safety limit* to the measured quantity.

References

[1] JCGM 100:2008 *(GUM) Evaluation of Measurement Data—Guide to the Expression of Uncertainty in Measurement* 1st edn (BIPM, IEC, IFCC, ISO, IUPAC, IUPAP, OIML—International Organization for Standardization)

[2] ISO/IEC 17025:2017 *General Requirements for the Competence of Testing and Calibration Laboratories* (Geneva: ISO)

[3] Ramsey M H, Ellison S L R and Rostron P (eds) 2019 *Measurement uncertainty arising from sampling: a guide to methods and approaches* 2nd edn (Gembloux: Eurachem) https://www.eurachem.org/images/stories/Guides/pdf/UfS_2019_EN_P2.pdf

[4] Magnusson B, Krysell M, Sahlin E and Näykki T 2020 *Uncertainty from sampling* 2nd edn (Taastrup: Nordtest) Nordtest Report TR 604 2020, www.nordtest.info

[5] JGCM 200:2012 *International Vocabulary of Metrology—Basic and General Concepts and Associated Terms (VIM)* 3rd edn (Sèvres: BIPM)

[6] IEC 60359:2001 *Electrical and electronic measurement equipment - Expression of performance* (Geneva: International Electrotechnical Commission (IEC))

Part B

Dealing with distributions

Chapter 11

The Normal distribution, the *t*-distribution, and the standard Normal distribution

Abstract: Among the various distributions in statistics, the Normal distribution is the most common and popular distribution. It finds many practical applications. The '*t*' and standard Normal distributions also have specific applications.

> *All models are wrong, but some are useful!*
> —George E P Box

It is well known that the underlying distribution in the GUM [1] is a Normal distribution. The data obtained by observations, the distribution of combined uncertainty, the central limit theorem (CLT), and the other assumptions in the GUM essentially revolve around the Normal distribution. Any substantial deviation from the Normal distribution amounts to a breach of the fundamental principles on which the edifice of the GUM is founded. The GUM also discusses the *t*-distribution (which is actually encountered in practical cases) at length. Let us study the fundamentals of these distributions and also understand what a standard Normal distribution is.

A typical Normal distribution and a *t*-distribution are shown in figure 11.1.

> *A statistician is someone who knows what to assume to be Gaussian.*
> —Dikran Marsupial

A Normal (Gaussian) distribution: The properties of a normal distribution are elaborated in chapter 29, 'Treating dominant non-Gaussian components' but repeated here for convenience.

A Normal distribution: Please refer figure 11.2.

    a. This is a unimodal distribution that has a bell-shaped curve and the same mean, mode, and median (which are measures of central tendency).

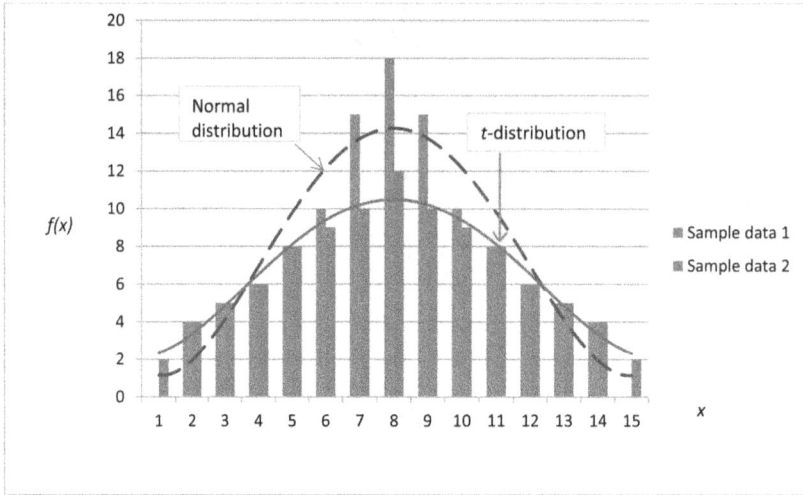

Figure 11.1. A Normal distribution and a *t-distribution*.

b. It is a symmetric continuous distribution. Due to this and (a) above, the areas under the curve on either side of the ordinate at $X = \mu$ are equal.
c. The x-axis is asymptotic to this distribution, which extends from $-\infty$ to $+\infty$. The significance of this property in measurements is that there is always some finite probability, however negligible, of getting an extreme value (also called an 'outlier') if we take an infinite number of measurements. (In practice, however, we confine our interest to a narrow interval of typically two standard deviations corresponding to a probability of approximately 95%.)
d. Its PDF can be defined by its mean μ and its standard deviation σ. (No additional parameters such as shape, location, etc. are required to define it. This is also a big advantage.)
e. Most of the data in economic and business statistics adhere to this distribution. It is also observed in many industrial, trade, and commercial applications and hence is a very realistic distribution. This characteristic makes it the most useful distribution. (It can be compared with a sinusoidal waveform, which has many applications in electrical engineering, mechanics, and motion studies, compared to other forms of waveforms.)
f. Many other distributions (continuous as well as discrete) approach the Normal distribution under limiting conditions, the significant condition being a large sample size (i.e. a sample size 'n' tending to ∞).
g. It is also obtained by convolving other distributions, albeit under some conditions. This characteristic is the basis of the CLT, which is the backbone of GUM.

The Normal distribution is defined by two parameters, namely the mean μ, and the standard deviation σ, which represent its central tendency and variability, respectively. For the same mean, the shape of a Normal distribution is more 'peaked' than

Table 11.1. The relation between level of confidence, standard deviation, and coverage factor.

Area under the curve/level of confidence	Standard deviation σ	Coverage factor k_p
50%	0.68σ	0.68
68.3%	1σ	1.0
95%	1.96σ	1.96
95.5%	2σ	2.0
99%	2.58σ	2.58
99.73%	3σ	3.0

that of a *t-distribution*; meaning it has less variability—a desirable property in any measurement. Its shape at the 'mode' is concave around the mean and becomes convex beyond the point of inflection. (The point of inflection (or inflection point) is a point on a curve at which the sign or concavity of the function changes.) If the area under the Normal distribution is taken to be 100%, the area within any interval of standard deviation 'σ' can be calculated. Typical values are given in table 11.1.

A *t*-distribution:

If z is a normally distributed random variable with an expectation $\mu(z)$ and standard deviation σ, and $\bar{z}$ is the arithmetic mean of n independent observations, with $s(\bar{z})$ the experimental standard deviation of $\bar{z}$, then the distribution of the variable $t = (\bar{z}-\mu(z))/s(\bar{z})$ is the *t*-distribution or Student's *t*-distribution, with degrees of freedom $v = n-1$. Essentially, a *t*-distribution is also a Normal distribution that is applied when the sample size is *small*. The shape of the *t*-distribution depends upon the degrees of freedom. The larger the tails (defined by larger kurtosis), the farther one needs to go from the mean to cover a given area. For example, for a given sample size, to cover 95% of the area in a *t*-distribution with four degrees of freedom, the interval must extend to 2.78 σ from the mean in both directions, compared to 1.96 σ for a Normal distribution. As the sample size increases (typically to more than 30), the value of $t_p(\infty)$ for a given fraction p equals the value of k_p in table 11.1 for the same p (see table 11.2 at the end of this topic).

Compared to a Normal distribution (that has the same area under the curve, for a specified interval):

- The area under the tails of a *t*-distribution is larger. This means, due to the limited sample size, that the probability of getting values (also called outliers) outside the specified interval is greater.
- It is flatter in the middle, meaning less 'peaked' (see figure 11.1) and tapers smoothly towards its tails (shows less inflection). In fact, this distribution is more realistic in routine measurements, as we have to settle for a limited sample size.
- Due to the above properties, for a smaller sample size, the *t*-factor becomes higher at all confidence levels, thereby giving a higher coverage factor and hence corresponding to higher expanded uncertainty. This is quite realistic, as with limited data we need to cover a 'larger' variance at a given level of confidence.

Table 11.2. Student's t-distribution for ν degrees of freedom. The t-distribution for ν defines an interval $-t_{p(\nu)}$ to $+t_{p(\nu)}$ that encompasses the fraction p of the distribution.

Degrees of freedom (ν)	t_p (for fraction p in %)					
	68.27	90	95	95.45	99	99.73
1	1.84	6.31	12.71	13.97	63.66	235.80
2	1.32	2.92	4.30	4.53	9.92	19.21
3	1.20	2.35	3.18	3.31	5.84	9.22
4	1.14	2.13	2.78	2.87	4.60	6.62
5	1.11	2.02	2.57	2.65	4.03	5.51
6	1.09	1.94	2.45	2.52	3.71	4.90
7	1.08	1.89	2.36	2.43	3.50	4.53
8	1.07	1.86	2.31	2.37	3.36	4.28
9	1.06	1.83	2.26	2.32	3.25	4.09
10	1.05	1.81	2.23	2.28	3.17	3.96
11	1.05	1.80	2.20	2.25	3.11	3.85
12	1.04	1.78	2.18	2.23	3.05	3.76
13	1.04	1.77	2.16	2.21	3.01	3.69
14	1.04	1.76	2.14	2.20	2.98	3.64
15	1.03	1.75	2.13	2.18	2.95	3.59
16	1.03	1.75	2.12	2.17	2.92	3.54
17	1.03	1.74	2.11	2.16	2.90	3.51
18	1.03	1.73	2.10	2.15	2.88	3.48
19	1.03	1.73	2.09	2.14	2.86	3.45
20	1.03	1.72	2.09	2.13	2.85	3.42
25	1.02	1.71	2.06	2.11	2.79	3.33
30	1.02	1.70	2.04	2.09	2.75	3.27
31	1.02	1.7	2.04	2.08	2.74	3.26
32	1.02	1.69	2.03	2.08	2.74	3.25
33	1.02	1.69	2.03	2.08	2.73	3.24
34	1.01	1.69	2.03	2.08	2.73	3.24
35	1.01	1.70	2.03	2.07	2.72	3.23
40	1.01	1.68	2.02	2.06	2.70	3.20
45	1.01	1.68	2.01	2.06	2.69	3.18
50	1.01	1.68	2.01	2.05	2.68	3.16
100	1.005	1.660	1.984	2.025	2.626	3.077
∞	1.000	1.645	1.960	2.000	2.576	3.000

A standard Normal distribution: If we take sample sets for different measurements, we obtain different means and sample standard deviations. Hence, the shape of the distribution for each set is different, and the calculation of the area for a given interval becomes complex. Further, a comparison of the distribution parameters becomes impossible. In order for a Normal distribution to be practically useful, it is necessary to have a single table. This is done by developing a standard Normal distribution, as shown in figure 11.2.

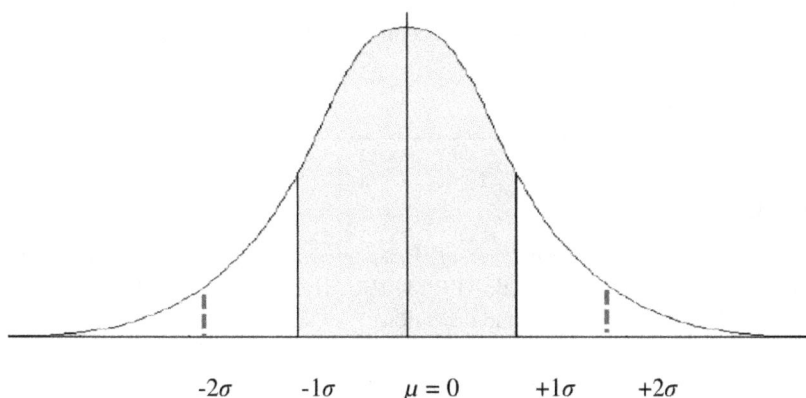

$$-2\sigma \qquad -1\sigma \qquad \mu = 0 \qquad +1\sigma \qquad +2\sigma$$

Figure 11.2. A standard Normal distribution.

A standard Normal distribution is obtained by normalizing the error of random variable x with respect to standard deviation. If x is a random variable under measurement, a new variable z can be defined as $z = (x-\mu)/\sigma$, where μ is the population mean or the expectation, and σ is the measurement standard deviation. The numerator is the measurement error; when divided by σ, z represents the normalized error. The purpose of standardization is to transform the random variable so that its distribution has zero mean and unit standard deviation, so as to be useful for comparison and analysis. A standard Normal distribution is characterized by $\mu = 0$ and $\sigma = 1$. Using standard Normal distribution tables, one can find the area between any interval that represents the level of confidence for that interval.

Points to ponder

From the above table, it can be observed that:

1. For a given p-value, as the degrees of freedom increase, the t-value decreases. The t-value is nothing but the coverage factor (k) that is used to multiply the combined standard uncertainty to obtain the expanded uncertainty. Thus, to get better (lower) expanded uncertainty, one must try to increase the degrees of freedom.

2. For given degrees of freedom, the t-value increases as the p-value increases. This means that at a higher level of confidence, the expanded uncertainty is greater for the given degrees of freedom. See chapter 16, 'Is a higher confidence level compatible with a larger expanded uncertainty?' for further interesting discussion.

3. If the value calculated (using the Welch–Satterthwaite formula) for the effective degrees of freedom is not an integer, the GUM advises that the degrees of freedom should either be interpolated or truncated to the next lower integer. It may, however, be noted that, at lower degrees of freedom (typically less than ten), a linear interpolation of the t-value may yield significantly different results. This is particularly true for levels of confidence (p-values) of 95.45% and above,

---

because the difference in t-values becomes nonlinearly greater for fewer degrees of freedom.

Reference

[1] JCGM 100:2008 *(GUM) Evaluation of Measurement Data—Guide to the Expression of Uncertainty in Measurement* 1st edn (BIPM, IEC, IFCC, ISO, IUPAC, IUPAP, OIML—International Organization for Standardization)

Chapter 12

Do Type A and Type B evaluations correspond to random and systematic errors?

Abstract: Usually, Type A method of evaluation is applied for evaluation of uncertainty of observed measurements. As the definition of random error contains phrases such as 'measurements under repeatability conditions,' some metrologists wrongly associate Type A evaluation with the evaluation of random error alone. Similarly, Type B evaluation is confused with the evaluation of systematic errors.

> *What we observe is not nature itself, but nature exposed to our method of questioning.*
>
> —Werner Heisenberg

As per the GUM [1], there are two types of evaluations of uncertainty, Type A and Type B.

The Type A evaluation of uncertainty is a method of evaluation by statistical analysis of a series of observations. Thus, for Type A evaluation of uncertainty, one needs to conduct an experiment in which one takes a few readings under repeatable conditions.

The Type B evaluation method is based on the evaluation of uncertainty by means other than the statistical analysis of a series of observations. Thus, here one can gather data and information in addition to, or in spite of, doing an experiment for a series of observations.

Typically, the information used for a Type B evaluation includes:
- previous measurement data,
- experience with or general knowledge of the behavior and properties of the relevant materials and instruments,
- the manufacturer's specifications,
- data provided in calibration and other certificates, and
- uncertainties assigned to reference data taken from handbooks.

After reviewing the literature on uncertainty, many laboratory personnel wonder whether the Type A and Type B evaluations are synonymous with random and systematic errors, respectively. To a larger extent and under normal measurement scenarios, there is a great deal of veracity in this equivalence. In routine measurement, we are not required to establish a systematic component by practical observations and treat it using the Type A method of evaluation. Measurement process variation is due to a combination of random and systematic influence factors. (Note that the terms 'random' and 'systematic' are valid for use with 'errors' and not uncertainties.) The random errors exist purely because of inherent, non-assignable, unpredictable causes which are supposed to be revealed by taking a large number of *independent* and repeat measurements. This idea is similar to the Type A evaluation described in the GUM. However, it is questionable whether the so-called *independent* measurements are really so, because there are many common stochastic, spatial, and temporal factors involved; system limitations as well as personal biases may result in insufficient independence. Thus, random variations arise because the influence quantities that can affect the measurement result cannot be held completely constant. Hence, despite careful control of the external influence factors, the variations due to random causes also vary, resulting in systematic errors that can either be corrected or accounted for. This is akin to the Type B evaluation described in the GUM.

Following a careful analysis of the evaluation of uncertainty, it was realized that this classification into random and systematic was inappropriate because all components of uncertainty have the same nature and can be treated identically. Hence, Recommendation INC-1 (1980) does not classify components of uncertainty as either 'random' or 'systematic.' It rather groups uncertainty components into two categories, 'Type A' and 'Type B,' based on their *method* of evaluation.

A systematic component (e.g. a correction) can be evaluated either by making a series of observations and treating them via the Type A method of evaluation or by using an assumed probability distribution and treating it via the Type B method of evaluation. Thus, it is the *method* and not the component itself that is distinguished in the GUM.

The note to clause 3.3.3 of the GUM clarifies that '*categorization of components of uncertainty (into random and systematic) can be ambiguous when generally applied. For example, a 'random' component of uncertainty in one measurement may become a 'systematic' component of uncertainty in another measurement in which the result of the first measurement is used as an input datum. Categorizing the methods of evaluating uncertainty components rather than the components themselves avoids such ambiguity. At the same time, it does not preclude collecting individual components that have been evaluated by the two different methods into designated groups to be used for a particular purpose.*'

Thus, it is the *method* and not the bizarre classification of components into 'random' and 'systematic' that is the premise of the GUM.

A fact one should know

Note that Type A and Type B are not the *types* of components but *methods* of evaluation of uncertainty. Clause 3.3.4 of the GUM clarifies this thus—'*...the classification is not meant to indicate that there is any difference in the nature of the components resulting from the two types of evaluation. Both types of evaluation are based on probability distributions.*' Hence, strictly speaking and by conception, there is no correspondence of the Type A method with uncertainties arising from random effects and no correspondence of the Type B method with uncertainties arising from systematic effects. The Type A and Type B categories apply to uncertainty and are not substitutes for the words 'random' and 'systematic.' Actually, in order to calculate the combined standard uncertainty, there is no need at all to classify uncertainty components as 'Type A' and 'Type B'! However, Recommendation INC-1 (1980) does stipulate the two *methods*, since such 'labels' are helpful for convenience, understanding, and discussion of ideas.

It should be noted that the GUM neither uses the terms 'random uncertainty' and 'systematic uncertainty' nor 'Type A error' and 'Type B error.' However, it does define and use the terms 'random error' and 'systematic error.' 'Uncertainty' and 'error' are completely distinct concepts, and this distinction is an important underpinning of the GUM uncertainty framework. (See chapter 1, 'Using correct terminology' for other topsy-turvy terms.)

Reference

[1] JCGM 100:2008 *(GUM) Evaluation of Measurement Data—Guide to the Expression of Uncertainty in Measurement* 1st edn (BIPM, IEC, IFCC, ISO, IUPAC, IUPAP, OIML— International Organization for Standardization)

IOP Publishing

A Practical Handbook on Measurement Uncertainty
FAQs and fundamentals for metrologists
Swanand Rishi

Chapter 13

What about reproducibility in uncertainty evaluation?

Abstract: For most practical measurements of physical measurands, repeatability is considered to be one component of uncertainty. But in many measurements, particularly in the food, biological, and chemical fields, reproducibility is considered instead of repeatability.

> *It doesn't matter how beautiful your theory is, it doesn't matter how smart you are. If it doesn't agree with experiment, it's wrong.*
>
> —Richard Feynman

One of the participants in a training course asked why reproducibility is not used in the Type A evaluation of uncertainty. Her argument was that reproducibility would be a more valid component, as it covers all possible combinations of variations! Good reproducibility certainly means good repeatability but not vice versa. 'The resulting uncertainty would be more realistic, representative, and incredibly robust!', she said.

The query is no doubt important. To address this interesting issue, let us look at clause 4.1.1 of the GUM [1], which deliberates the Type A evaluation of standard uncertainty. It calls for n independent observations to be obtained under the *same* conditions and gives reference of clause B.2.15. Variations in *repeat* observations that we observe while using Type A method of evaluation are assumed to arise because the influence quantities that affect the measurement result cannot be held completely constant. Thus, we make singular efforts to keep these factors as constant as possible and also try to avoid or mitigate their interference.

Clause B.2.15 of the GUM is about 'repeatability' and defines it as 'closeness of the agreement between the results of successive measurements of the same measurand carried out under the *same* conditions of measurement.' It is worth noting that the next clause, B.2.16, gives a definition of 'reproducibility,' but it is not referenced by the GUM in any of the discussions about the Type A evaluation of uncertainty. Reproducibility is the 'closeness of the agreement between the results of measurements of the same measurand carried out under *changed* conditions of measurement.'

Traditionally, the evaluation of uncertainty was based on the concept of probability, which was viewed as applicable only to the events that can be repeated a large number of times *under essentially the same conditions*. This points to nothing but repeatability. (Unfortunately, this *relative frequency*-based concept was so ingrained in the old school of statisticians that it was difficult for them to accept the concept of combining variances obtained from *a priori* distributions with those from *frequency-based* distributions.)

We can also refer to GUM clause B.2.21 for random error (which is supposed to be obtained by Type A evaluation), which specifically uses the words 'repeatability conditions.' (Even clause B.2.22 that defines systematic error uses the same words).

It is also well known that calibration laboratories try to estimate uncertainty by exercising the best possible control over external influence factors. The same efforts are also made to get the lowest calibration and measurement capability (CMC), as that is *the* measure of a laboratory's capability. (The discussion about repeatability above is more relevant in case of calibration measurements.)

In the realm of reproducibility, many measurement conditions such as the equipment, method, operator, location, time, etc. have to be changed. This is bound to result in more variation and ultimately a poorer or wider CMC. The way to combine uncertainties obtained under varied conditions of reproducibility is again a moot question. The most forbidding aspect is that it becomes impossible to incorporate this uncertainty into another result, as this would amount to comparing apples with oranges; the evaluation conditions in the domain of reproducibility have many possible combinations. Further, it becomes difficult to participate in interlaboratory comparison (ILC) programs, which are necessarily based on a protocol that mandates similarity in measurement conditions.

Facts one should know

FACT 1: A two-operator comparison could be a good way to meet the requirements of clause 7.2 of ISO/IEC 17025:2017 [2] for the 'Selection, verification and validation of methods.' In the realm of reproducibility, any one condition of the measurement process can be changed at a time. Thus, it can be effectively employed to evaluate an operator's knowledge and skill, as it is possible for a laboratory to keep all the other conditions the same. The E_n value could be used as a criterion for their evaluation.

FACT 2: Due to the limitations of the GUM as regards its mathematical modeling approach, alternate approaches are actively being used for the evaluation of uncertainty, particularly in analytical testing and quantitative testing. In some such methods, reproducibility is taken into account to evaluate the uncertainty of quantitative test measurements. In these tests, the 'reproducibility standard deviation' is considered to be a proper estimate for the evaluation of measurement uncertainty, since repeatability standard deviation does not include major contributions. See chapter 34, 'Alternative approaches in uncertainty evaluation.'

References

[1] JCGM 100:2008 *(GUM) Evaluation of Measurement Data—Guide to the Expression of Uncertainty in Measurement* 1st edn (BIPM, IEC, IFCC, ISO, IUPAC, IUPAP, OIML—International Organization for Standardization)

[2] ISO/IEC 17025:2017 *General Requirements for the Competence of Testing and Calibration Laboratories* (Geneva: ISO)

Chapter 14

How is it that we can combine different distributions?

Abstract: It is necessary to provide a single value of uncertainty for a measurement result. The uncertainty components evaluated by the Type A and Type B methods of evaluation usually have different (assumed) distributions. Combining them is like mixing apples with oranges. How is it done, then?

Prediction is very difficult, especially if it's about the future.

—Niels Bohr

We evaluate uncertainty by the Type A and Type B methods separately, usually using repeatability and other sources, respectively. Thus, we obtain two quantities of uncertainty in terms of standard deviations. These are not useful for comparison with other results of uncertainty, which are generally in the form of a single quantity. Recommendation INC-1 (1980) states that combined standard uncertainty shall be obtained by combining variances by the usual method (read root sum square(RSS)) and expressed in the form of a standard deviation.

All the components are assumed to have a unique probability distribution and are required to be combined (convolved) together to obtain the combined uncertainty. In the vast majority of cases, a Normal distribution from the Type A method and a few uniform distributions from the Type B method are required to be combined. This is something like putting a few oranges and cherries (Type B components with different shapes of distributions) into a basket of apples (Type A components with an assumed Normal distribution) and declaring it to be a basket of 'homogeneous' apples (and charging for apples)!

How is it possible to combine different distributions of uncertainty components for uncertainty evaluation? It is the central limit theorem (CLT) that comes to our rescue. This theorem states that the distribution of measurand Y is *approximately normal* if the input quantities Xi are (a) independent and (b) the variance of Y is much larger than variance of any single component from a non-normally distributed X_i. Simply put, it states that even if the distribution of input variables is not Normal, the distribution of the *sample means* (output estimate) always approaches the Normal distribution as the sample size increases without bound. The CLT makes it possible to assign a confidence level in terms of probability to the combined uncertainty. Further, it implies that:

(i) the convolved distribution converges towards the Normal distribution as the number of input quantities increases;

(ii) the convergence is more rapid when the values are closer to each other; and

(iii) the closer the distributions of the input quantities to the Normal distribution, the fewer of them are required to yield a Normal distribution for the output quantity.

(In our analogy with a basket of fruit, the basket is *almost* full of apples, with very few oranges and cherries!)

According to the CLT, if the input quantities are fairly close to each other, the convolution approaches the Normal distribution faster. That is to say, the combined standard uncertainty $u_c(y)$ should not be dominated by a standard uncertainty component obtained from either a Type A evaluation based on just a few observations or by a standard uncertainty component obtained from a Type B evaluation based on an assumed non-normal (generally rectangular) distribution.

The convolution approaches closer and closer to the Normal distribution as the input quantities (n) increase, irrespective of the shape of their distributions. Ideally, as $n \sim \infty$, the convolved distribution becomes very close to normal. In practice, if three or more distributions of *similar* magnitude are present, they combine to form a reasonable approximation to the Normal distribution.

If the input quantities are closer to normal, fewer quantities are required for the convolved (i.e. resultant) distribution to be close to normal, and the linear approximation implied by the 'law of propagation of uncertainty' is adequate.

However, in practice, the distributions of input quantities are usually estimates, not quantities themselves. (In other words, we deal with sample data rather than population data.) It is also unrealistic to expect a high level of confidence that can be assigned to a given interval of the estimate. The process of convolving probability distributions is quite complex and, as such, their convolutions are rarely implemented. Hence, in routine evaluations of uncertainty, approximations (i.e. *estimates* of input quantities and *not distributions*) are used that take advantage of the CLT.

The practical significance of the CLT is that if combined uncertainty is not dominated by any single component of the input quantity (Type A with very few readings or Type B with assumed uniform distribution), the coverage factor k_p can be obtained from a Normal distribution corresponding to a confidence level p. Because of this theorem, it is sufficient to assume that the resultant probability distribution of the output is a t-distribution, so that we can take $k_p = t_p(\nu_{\mathrm{eff}})$, with Student's t-factor, t_p, based on effective degrees of freedom ν_{eff} obtained from the Welch–Satterthwaite formula.

This process of combining different distributions is also called 'convolution.' The trapezoidal distribution is equivalent to the convolution of two rectangular distributions: one has a half-width a_1 equal to the mean half-width of the trapezoid, $a_1 = a(1 + \beta)/2$, the other has a half-width a_2 equal to the mean width of one of the triangular portions of the trapezoid, $a_2 = a(1—\beta)/2$. The convolution of three uniform distributions of the same half-width results in a near-Normal distribution that has a 95% interval of 1.937σ, compared to a corresponding interval of 1.96σ for a Normal distribution. Note that this amounts to a deviation (underestimation) of just 1.5% from the Normal distribution.

Thus, the CLT is the backbone of the uncertainty evaluation process. However, it is important to ensure that its conditions are fulfilled. (See chapter 32, 'Analyzing the results.')

Facts one should know

It may be noticed that in most uncertainty evaluations, uniform and Normal distributions are assumed virtually all the time. In such cases, a rule of thumb that can be used to check the dominance of a rectangular uncertainty component w.r.t. u_c is as follows:
1. If the standard uncertainty of a component is more than 1.4 times the combined standard uncertainty of the remaining components, it is dominant and needs to be treated differently.
2. If it is not, then the coverage factor k is within 5% of the customary value of 2.00 at a 95.45% confidence level at effective degrees of freedom $\sim \infty$; hence, the usual method of evaluation should be followed.

However, the GUM [1] permits the use of other analytical methods when the conditions of the CLT are not met, resulting in unacceptable uncertainty (e.g. the convolution of a Normal distribution with a rectangular distribution or other distributions).

Reference

[1] JCGM 100:2008 *(GUM) Evaluation of Measurement Data—Guide to the Expression of Uncertainty in Measurement* 1st edn (BIPM, IEC, IFCC, ISO, IUPAC, IUPAP, OIML—International Organization for Standardization)

Chapter 15

Guidelines for the selection of triangular and trapezoidal distributions

Abstract: Triangular and trapezoidal distributions are assumed for the evaluation of the standard uncertainty of certain components. However, there are no specific objective guidelines for these assumptions. This chapter offers some objective criteria with which to make these assumptions.

> *I couldn't claim that I was smarter than sixty-five other guys—but the average of sixty-five other guys, certainly!*
>
> —Richard Feynman

The most common probability distributions discussed in most of the national standards/guides are the normal, uniform (or rectangular), trapezoidal, triangular, and U (or arcsine) distributions. However, it may be observed that in most uncertainty evaluations, uniform and Normal distributions are assumed more than 90% of the time. The Normal distribution is valid only in two very specific cases: one in repeatability (Type A) and other in uncertainty based on a calibration certificate (Type B).

We will examine some distributions related to this topic:

1. **A uniform distribution:** Figure 15.1 shows this distribution. This is one of the most popular distributions assigned to Type B components. The reasons cited for its popularity are: (a) ease of application, (b) the recommendations made in the GUM [1], and (c) *safer* uncertainty. These are discussed in short below.

   a. **Ease of application**—this distribution is usually assumed when an error component is bound by upper (+) and lower (−) limits and many of the error components have such limits. For a uniform distribution, the standard uncertainty is obtained by simply dividing the estimated limits of error by $\sqrt{3}$. Laboratory auditors are perhaps comfortable with this

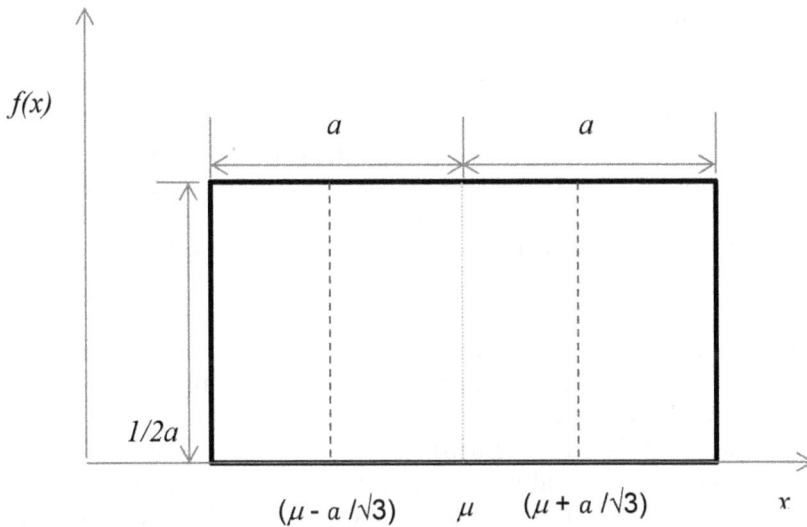

Figure 15.1. A uniform or rectangular distribution. Reproduced from [5] with permission from NCSL International. © 2022 NCSL INTERNATIONAL. ALL RIGHTS RESERVED.

method, and its widespread use may be due to the fact that the tools required for Type B uncertainty evaluation were not available when the GUM was originally published. However, this argument is no longer justifiable, as many computational tools are now available. As stated by NASA-HDBK 8739.19-3: 2010 [2] in Clause B.12.3, '*Since the introduction of the GUM, methods have been developed to systematically and rigorously apply distributions that are physically realistic.*'

b. **The recommendations of the GUM**—the proponents of the uniform distribution cite the GUM in support of its ad-hoc use. This argument is also nonviable, as most of the GUM approach presupposes that the underlying error distribution is normal. Clause 4.3.7 of the GUM [3] states that '*If there is no specific knowledge about the possible values of X_i within the interval, one can only assume that it is equally probable for X_i to lie anywhere within it (a uniform or rectangular distribution of possible values).*' Taking refuge in a lack of specific knowledge cannot be an excuse these days, due to the ready availability of software tools for data analysis.

c. **Safer uncertainty**—the use of the uniform distribution results in larger uncertainties than those of the trapezoidal or triangular distributions, which is obvious from their formulae. Even though the goal of many laboratories during accreditation audits is to demonstrate the smallest possible measurement uncertainties, it is often considered safe and hence conservative for a laboratory to assume a uniform distribution. This viewpoint ignores potential benefits and does not seem fit for pragmatic metrologists. This safe approach is also deprecated in GUM appendices E1 and E2.

Many standards/guides and documents caution against the widespread use of a uniform distribution, e.g. Clause B.12 of NASA HDBK 8739.19-3:2010 [2], clause 4.3.9 of the GUM and ANSI Z540.2 [4], clause A.2.1 of NABL 141 [3], and clause 4.6 of NIST TN 1297 express reservations about the liberal use of the uniform distribution and call upon metrologists to be realistic about its assumption.

2. **A trapezoidal distribution:** This distribution results from the convolution of two uniform distributions that have the same mean and variance. It can also be assumed when data are interpolated over a small range and its applicability can be verified from actual data.

   Figure 15.2 shows a symmetric trapezoidal distribution with a base width of $2a$ and an upper width of $2a\beta$, where $0 \leqslant \beta \leqslant 1$ (β is the ratio of the top width to the base width). This distribution applies when the values are spread through the central portion of the interval with equal probability. The probability gradually reduces towards the extremes of the base width, beyond which it is zero.

   This distribution is more realistic than a uniform distribution, because the probability that the values lie in some middle portion is higher than the probability that they occur uniformly throughout the lower and upper limits (especially near the bounds). Further, there is no step change of function; the probability gradually decreases from a uniform value to lower values until it reaches zero. By establishing β, a uniform distribution can be transformed into a trapezoidal distribution.

3. A triangular distribution: As shown in figure 15.3, a triangular distribution is a lower limiting case of a trapezoidal distribution with $\beta = 0$. The terminologies used for its selection in some documents/guides (NABL

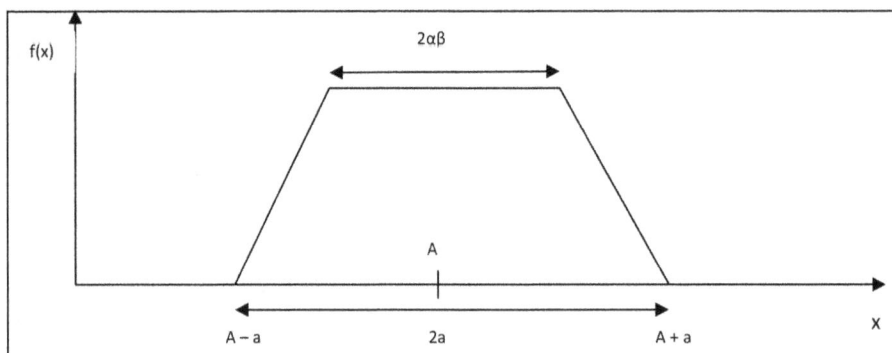

Figure 15.2. A symmetric trapezoidal distribution. Reproduced from [5] with permission from NCSL International. © 2022 NCSL INTERNATIONAL. ALL RIGHTS RESERVED.

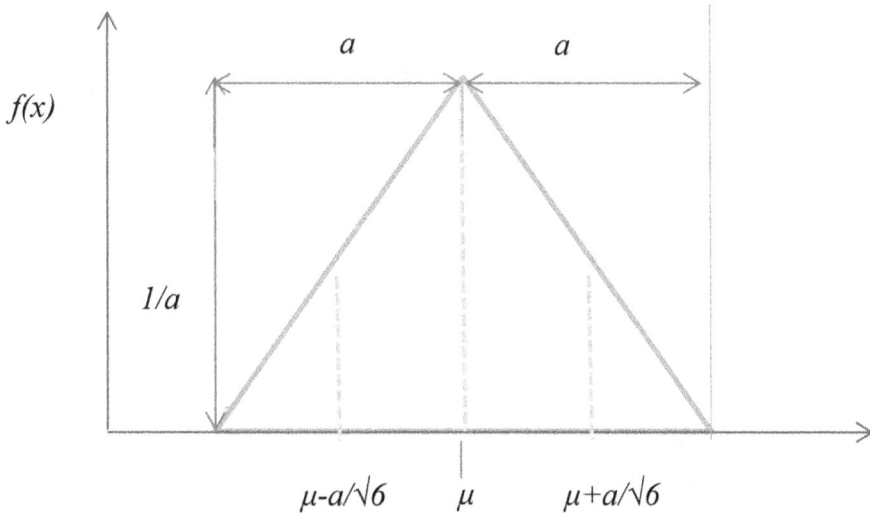

Figure 15.3. A triangular distribution. Reproduced from [5] with permission from NCSL International. © 2022 NCSL INTERNATIONAL. ALL RIGHTS RESERVED.

[3], NATA MU book [7], NIST TN 1297 [6]) are subjective ('in the middle range,' 'near the center,' etc) and hence predisposed to dispute or disagreement. Hence, there is a need for some objective criteria. With careful analysis of the data, one can select a triangular or trapezoidal distribution in place of a uniform distribution.

The method of data analysis was presented in my paper that was published in the March 2013 issue of NCSLI Measure titled 'Proposed Guidelines for the Selection of Trapezoidal and Triangular Distributions for an Uncertainty Evaluation' [5]. (I do not delve into the details of the proposal here. Interested metrologists may refer the paper.) As a perfect triangular or uniform distribution is highly improbable in reality, the proposed objective criteria for their selection based upon the value of β are given below:

- use a triangular distribution when $\beta < 0.3$;
- use a trapezoidal distribution when $0.3 \leqslant \beta \leqslant 0.65$;
- use a uniform distribution when $\beta > 0.65$.

The use of a trapezoidal or triangular distribution in place of a uniform distribution effectively reduces the uncertainty contribution of that particular component. It is necessary to analyze the data over a sufficient period for the value of β. Certainly, some efforts are required to establish β, but given the various statistical tools at one's disposal today, this should not be too resource intensive. It is likely that in today's competitive world, even a small reduction in uncertainty in high-precision measurement would reap remarkable advantages.

Figure 15.4 shows the three distributions as per the proposed guidelines with respect to β, considering a base width (specification limit) of 1 (100%).

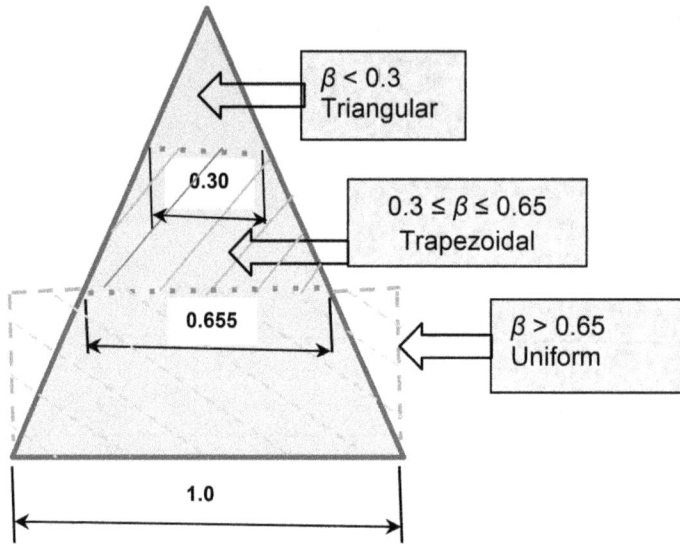

Figure 15.4. The three distributions as per the proposed guidelines with respect to β. Reproduced from [5] with permission from NCSL International. © 2022 NCSL INTERNATIONAL. ALL RIGHTS RESERVED.

References

[1] JCGM 100:2008 *(GUM) Evaluation of Measurement Data—Guide to the Expression of Uncertainty in Measurement* 1st edn (BIPM, IEC, IFCC, ISO, IUPAC, IUPAP, OIML— International Organization for Standardization)

[2] NASA-HDBK-8739.19-3 2010 *Measurement Uncertainty Analysis Principles and Methods, NASA Measurement Quality Assurance Handbook - ANNEX 3* (Washington, DC: NASA)

[3] NABL 141 2020 *Guidelines for Estimation and Expression of Uncertainty in Measurement Issue 4* (Gurugram: NABL India)

[4] ISO/IEC 17025:2017 *General requirements for the competence of testing and calibration laboratories* (Geneva: ISO) (ANSI/NCSL Z540.2-1997 was superseded by ANSI/NCSL Z540.3-2006. ANSI/NCSL Z540.3-2006 is also withdrawn and superseded by ISO/IEC 17025:2017.)

[5] Rishi S 2013 Proposed guidelines for the selection of trapezoidal and triangular distributions for an uncertainty evaluation *J. Meas. Sci.* **8** 72–7

[6] NIST 1994 *Guidelines for Evaluating and Expressing the Uncertainty of NIST Measurement Results* (Gaithersburg, MD: NIST) NIST Technical Note 1297, https://www.nist.gov/pml/nist-technical-note-1297

[7] Cook R R 2002 *Assessment of Uncertainties of Measurement for Calibration and Testing Laboratories* 2nd edn (Rhodes, NSW: NATA) [No longer available on NATA Website]

Chapter 16

Aren't higher confidence level and larger uncertainty contradictory?

Abstract: When a higher level of confidence is taken (with a corresponding higher coverage factor), one gets a higher value of expanded uncertainty. One may feel that at a higher level of confidence, one must get a lower expanded uncertainty, which is desirable. But is this contradictory?

Statistics can be made to prove anything—even the truth.

One of the peculiar issues that baffles many metrologists (particularly beginners) is explaining why uncertainty becomes larger at a higher confidence level. Their ingenuousness is certainly understandable, as one naturally thinks that when a person has a higher confidence level about some event to occur, he or she is surer about its favorable outcome. Therefore, the chances of a favorable event failing to occur should be less at higher confidence levels. In the case of uncertainty, when we estimate it at a higher confidence level, we are actually making it look poorer and more unfavorable. With wider uncertainty, the laboratory drops further down the measurement capability chain. So, why not claim uncertainty at a lower confidence level?

To answer this apparently perplexing query, let us look at it from a different perspective using an example. Suppose a college girl is asked about the percentage mark she expects to get in an exam that she has recently given. Let us say her answer is 'between 80 and 90.' When further asked how sure she is, she would probably say 90%. If asked 'with what confidence level will the marks be between 83 and 87 and between 84 and 86,' she would be bewildered, as the range of expected marks has been tightened. (It is tightened equally from the original range for moderation and to draw parallels with uncertainty.) 'With a smaller range, I need to taper down to some extent,' she muses. Assuming that she is rational enough, she may ponder over and by applying common sense may arrive at confidence levels of, say, 80% and 70%, respectively. If that range of expected marks were to be further narrowed to 84.5%–85.5%, she would be at a loss. She would probably say: 'it is not possible to quantify, but my confidence would be certainly much lower than 70%.' And finally, if asked about her confidence level for getting exactly 85%, she would be left high and dry! Thus, in this example, we are moving from an *interval* estimate to a *point* estimate, progressively making it increasingly difficult to quantify the confidence level. On the other hand, if we enlarge the range of marks so that it lies from 0% to 100%, anyone can claim a 100% confidence level! Thus, the narrower the range, the lower the confidence level, and the wider the range, the higher the confidence level.

From the above example, it is clear that our confidence level continues to reduce as the band or range of expectation contracts and vice versa. Thus, in order to accommodate a wider range of expected outcomes, we have to enlarge the confidence level. This is also clear from the shape of the Normal distribution, which is supposed to be the resultant distribution of combined uncertainty. If we want to factor in the probability of some very rare events (or less probable readings in the case of measurement), the interval in terms of standard deviation has to be widened. (And, as everyone knows, this standard deviation and coverage factor k are the same if the effective degrees of freedom are infinite.)

Thus, instead of saying that 'at a higher confidence level the uncertainty gets larger and poorer,' it is more apt to say 'in order to accommodate greater uncertainty, the confidence level has to be higher.'

Facts one should know

FACT 1: It is worth noting that even though uncertainty is lower at a lower confidence level, its 'absolute' or 'usable' value—which is 'combined uncertainty'—remains the same. In Recommendation 1 (CI 1986) [1], the CIPM requested that the 'combined standard uncertainty u_c be used by all participants in giving the results of all international comparisons or other work done under the auspices of the CIPM and Comités Consultatifs.' So, if two laboratories claim their uncertainties at different confidence levels, one of them claiming lower uncertainty at 90% and the other at recommended level of 95%, their absolute uncertainties can be compared only by dividing those figures by the corresponding coverage factors k of 1.65 and 1.96, respectively (thus, obtaining the combined uncertainty). Using the law of propagation of uncertainty, we can incorporate only the *combined* uncertainty (and not the expanded uncertainty) of one result into another result. Thus, practically speaking, there is no advantage in, so to say, *'acquiring'* a lower uncertainty by adopting a lower confidence level.

FACT 2: Reproduced below is clause 6.2.2 from the GUM [2], which clarifies some terms involving the word 'confidence.'

*'The terms **confidence interval** and **confidence level** have specific definitions in statistics and are only applicable to the interval defined by U when certain conditions*

are met, including that all components of uncertainty that contribute to $u_c(y)$ be obtained from Type A evaluations. Thus, in this Guide, the word 'confidence' is not used to modify the word 'interval' when referring to the interval defined by U; and the term 'confidence level' is not used in connection with that interval but rather the term 'level of confidence.' Hence GUM uses the term 'level of confidence' rather than 'confidence level.' More specifically, U is interpreted as defining an interval about the measurement result that encompasses a large fraction p of the probability distribution characterized by that result and its combined standard uncertainty, and p is the coverage probability or level of confidence of the interval.'

The NIST/SEMATECH Engineering Statistics Handbook [3] clause 7.1.4 elaborates thus: *'Confidence intervals are constructed at a confidence level, such as 95%, selected by the user. What does this mean? It means that if the same population is sampled on numerous occasions and interval estimates are made on each occasion, the resulting intervals would bracket the true population parameter in approximately 95% of the cases. A confidence stated at a $1-\alpha$ level can be thought of as the inverse of a significance level, α.'*

The GUM and NIST TN 1297 [4] use the term 'level of confidence' rather than 'confidence level,' while many other guides use the latter term.

References

[1] CIPM 1986 *International Committee for Weights and Measures* (Sèvres: BIPM)

[2] JCGM 100:2008 *(GUM) Evaluation of Measurement Data—Guide to the Expression of Uncertainty in Measurement* 1st edn (IPM, IEC, IFCC, ISO, IUPAC, IUPAP, OIML— International Organization for Standardization)

[3] NIST/SEMATECH 2012 *e-Handbook of Statistical Methods* (Gaithersburg, MD: NIST)

[4] NIST 1994 *Guidelines for Evaluating and Expressing the Uncertainty of NIST Measurement Results* (Gaithersburg, MD: NIST) NIST Technical Note 1297, https://www.nist.gov/pml/nisttechnical-note-1297

IOP Publishing

A Practical Handbook on Measurement Uncertainty
FAQs and fundamentals for metrologists
Swanand Rishi

Chapter 17

Mind the correlations

Abstract: The law of propagation of uncertainty (LPU) and the central limit theorem (CLT) are applicable under certain conditions. Metrologists should be aware of these conditions, one of them being correlation between input quantities.

> *All too often when liberals cite statistics, they forget the statisticians'*
> *warning that correlation is not causation.*
>
> —Thomas Sowell

The whole ambit of the GUM [1] is based on the LPU and the CLT. However, these theorems are applicable under certain conditions. The details of these are discussed elsewhere in chapters such as chapter 7, 'The law of propagation of uncertainty' and chapter 14, 'How is it that we can combine different distributions?' The LPU in its simplest form is valid if all input quantities are independent or uncorrelated; otherwise, correlation must be taken into account by way of a correlation coefficient.

Identifying correlation: Clause C.2.8 of the GUM defines correlation as 'the relationship between two or several random variables within a distribution of two or more random variables.' Thus, two quantities are correlated if they have been measured simultaneously or if they are determined from common observations or by using a common instrument. Correlation usually exists due to effects of common influence quantities such as temperature, humidity, gravitational acceleration, and barometric pressure. Often, correlation exists, but its extent is insignificant. Hence, experience and insight are essential when deciding the extent of correlation. A few examples are:

(a) The measurement of the temperature of two quantities is performed using a common thermometer.

(b) The elongation strength of a thread and the humidity are observed simultaneously (they have a strong correlation).

(c) The capacitance and the reactance of a capacitor are determined from simultaneous observations.

(d) Humidity is calibrated using the dry and wet bulb method.

In all such and similar cases, the covariances and correlation coefficients should be calculated. If correlation exists, the covariance can be established experimentally (Type A) or using *a priori* knowledge (Type B). The former method is expensive but more accurate. When reporting the result, the GUM recommends that the estimated covariance or correlation coefficient of each input quantity should be given, and if it is close to one, it should preferably be given to three significant digits.

Covariance usually has an inconvenient dimension and magnitude and is hence difficult to grasp, whereas correlation coefficient is a pure number between -1 and $+1$ (both inclusive) and hence easy to comprehend.

17.1 The correlation coefficient

The correlation coefficient r is a measure of the relative mutual dependence of two variables x_i and x_j, equal to the ratio of their covariances to the positive square root of the product of their variances. Mathematically,

$$r(x_i, x_j) = \frac{u(x_i,x_j)}{u(x_i)u(x_j)}, \tag{17.1}$$

where $u(x_i)$ and $u(x_j)$ are the uncertainties of the (correlated) input estimates.

The covariance is given by

$$u(x_i,x_j) = \frac{1}{n-1} \sum_{i=1}^{n}(x_i - \bar{x}) \ (x_j - \bar{x}). \tag{17.2}$$

Some important points to note about the correlation coefficient are:
1. If two random variables are independent, their correlation coefficients (and covariance also) are zero, but the converse is not necessarily true.
2. In comparison measurements, the unit under test (UUT) and the standard usually have the same or similar nominal values. If α is the ratio of the UUT measurement and the standard value and $u(\alpha)$ is the uncertainty of comparison, the correlation coefficient is given by

$$r_{ij} = \left[1 + \left(\frac{u(\alpha)}{u(R_s)/R_s}\right)^2\right]^{-1}. \tag{17.3}$$

In equation (17.3), $u(R_s)$ is the uncertainty of the standard R, and R_s is its quantity value.

If the comparison measurement method/procedure is well characterized, $u(\alpha)$ is negligible compared to the relative uncertainty of the standard (a common case). This results in r_{ij} close to one. In such cases, the uncertainty of UUT is the *same* as that of the standard.

3. If all input estimates are correlated with a correlation coefficient close to +1 ($r = +1$), the combined uncertainty reduces to

$$u_c(y) = \sum_{i=1}^{n} \frac{\partial f}{\partial x_i} u(x_i). \qquad (17.4)$$

Thus, when $r = +1$, the combined uncertainty is the *linear sum* of the uncertainties of each estimate. Hence, if each input estimate in a sample of size n has the *same* uncertainty, the combined uncertainty with correlation is equal to $\sqrt{n}$ times the uncertainty obtained under the uncorrelated assumption!

This is a typical measurement situation in which a number n of the same UUTs (by value) are used in conjunction. For example:
- Ten resistors of 1 Ω each are compared with a standard resistor of 1 Ω and then connected in series to scale up the value to 10 Ω.
- A weighing balance is to be calibrated at 5 kg with 5 weights of 1 kg each that were calibrated against a standard weight of 1 kg.

4. A negative correlation coefficient reduces the uncertainty!
5. If the input estimates x_i and x_j are correlated and if a change δ_i in x_i produces a change δ_j in x_j, then the correlation coefficient $r(x_i, x_j)$ associated with x_i and x_j can be approximately estimated by

$$r(x_i, x_j) \approx u(x_i)\delta_j / \left[u(x_j)\delta_i \right], \qquad (17.5)$$

where $u(x_i)$ and $u(x_j)$ are the respective uncertainties.

This is useful for estimating correlation coefficients experimentally. It can also be used to calculate the approximate change in one input estimate due to a change in another (input estimate) if their correlation coefficient r is known.

6. The variances and correlation coefficients are associated with the *errors* of the input quantities rather than with the input quantities themselves.

A fact one should know

It shall be noted that covariance shows only the *mutual dependence* of two input random variables, while the correlation coefficient is a measure of the *relative* mutual dependence of two variables. Thus, it shows the strength of the relationship between the two variables, whether positive or negative. A zero correlation coefficient means there is no relationship or dependence between the two variables.

The following example illustrates the calculation of the covariance and the correlation coefficient.

Let X and Y be two input quantities that are observed simultaneously and are therefore correlated. Five sets of readings are shown in table 17.1.

Table 17.1. The calculation of the covariance and the correlation coefficient for inputs X and Y.

Set No.	X	Y	$X_i - \overline{X}$	$Y_i - \overline{Y}$	$(X_i - \overline{X}) \times (Y_i - \overline{Y})$
1	15.223	29.241	0.0274	0.020	0.0005
2	14.587	29.362	−0.6086	0.141	−0.0858
3	14.964	29.552	−0.2316	0.331	−0.0766
4	15.541	28.981	0.3154	−0.240	−0.0829
5	15.663	28.969	0.4674	−0.252	−0.1178
Mean	$\overline{X} = 15.595$	$\overline{Y} = 29.221$			
Standard Deviation, SD	SD_X = 0.437	SD_Y = 0.250			
SUM	$= \sum (X_i - \overline{X}) \times (Y_i - \overline{Y})$				−0.3626
Covariance	$= \sum (X_i - \overline{X}) \times (Y_i - \overline{Y})/(5{-}1)$				−0.0907
Correlation coefficient, r	$=$ Covariance/(SD_X) $\times$ (SD_Y) $= -0.0907/\ (0.437 \times 0.250)$				−0.83

Note: figures rounded in some places; r is usually given at up to two significant digits.
Note: Excel has various functions to calculate the SD, the covariance, and the correlation coefficient.

Thus, a correlation coefficient r of −0.83 shows a strong negative correlation between inputs X and Y.

Facts one should know

FACT 1: One can find a close resemblance between the 'correlation coefficient' given by equation (17.5) and the 'sensitivity coefficient.' (See chapter 25, 'What is the significance of the sensitivity coefficient?'.) However, there is a difference: the former indicates a relationship between two *input* estimates, whereas the latter indicates a relationship between an *input* and an *output* estimate.

Note that the correlation is invalidated if the two quantities in question are not corrected for the influencing effect and the effect is *separately* reported in terms of additional input quantities with their standard *uncertainties*. The decision whether to apply correlation or incorporate the effect separately is driven by cost and benefit.

FACT 2: A strong correlation does not necessarily imply a causal link between the two variables. For example, if we find in a sample survey that the height of a person and his/her typing speed have a correlation coefficient of 0.68, it does not mean that a greater height increases one's typing speed!

FACT 3: The correlation coefficient r is a measure of *linear* relationship that indicates the mutual dependence of two variables. Thus, $r = 0$ does not mean there is no relationship between these variables, rather it means there is zero *linear* relationship.

For example, there could be a quadratic or a strong curvilinear (nonlinear) relationship between these variables, but their correlation coefficient r could be zero.

The assumptions underlying the validity of the correlation coefficient are:
1. The two variables are random samples.
2. At least one of the two variables has a Normal distribution in the population.

Although the GUM deprecates the use of 'true' values and departs from the earlier 'error' model of uncertainty, by interpreting the LPU in terms of 'true' value and 'error,' it proves that the variances and correlation coefficients are associated with the *errors* of the input quantities rather than with the *input quantities* themselves.

Reference

[1] JCGM 100:2008 *(GUM) Evaluation of Measurement Data—Guide to the Expression of Uncertainty in Measurement* 1st edn (BIPM, IEC, IFCC, ISO, IUPAC, IUPAP, OIML—International Organization for Standardization)

Part C

Sample size and analysis

IOP Publishing

A Practical Handbook on Measurement Uncertainty
FAQs and fundamentals for metrologists
Swanand Rishi

Chapter 18

Sampling distributions

Abstract: The process of evaluation of uncertainty is part of what is called 'inferential statistics,' in which an inference about a large amount of data—called the *population*—is drawn from a small amount of data—called the *sample*. Naturally, the conclusion drawn from the measurement results depends upon the sample characteristics and the distribution of the sample or the *sampling distribution*. Sampling distributions help us draw inferences about a population.

> *A little is as much as a lot, if it is enough!*
>
> —Steve Brown

The GUM requires the evaluation of uncertainty to be done using the Type A and Type B methods of evaluation. For Type A evaluation, we need to take readings of the quantity under measurement. These readings have to be taken on sample basis and are supposed to represent the population or the universe. Through Type A evaluation, we find the mean value, which is supposed to be the best estimate of the quantity. Based on Type A and Type B standard uncertainty, we predict a possible range or spread of the best estimate of the measurand.

So, this process of evaluation of uncertainty is part of what is called 'inferential statistics.' Inferential statistics is defined as statistics in which an inference about a large amount of data—called the *population*—is drawn from a small amount of data —called the *sample*. Naturally, the conclusions depend upon how well the sample represents the population, the sampling technique used, the skill of the operator, the sample preparation methods (where applicable), etc. Apart from these, the result also depends upon the distribution of the sample or *sampling distribution*. Sampling distributions help us draw inferences about a population based on the study of the sample. Usually, the samples are selected by random sampling methods to avoid bias in sampling.

18-1

A fact one should know

There are two types of populations: finite and infinite.

A population that has a finite number of items/data is called a finite population; examples include the number of people staying in a state or a number of cars in a city. It is thus *discrete* (countable) in nature.

A population that has an infinite number of items/data is called an infinite population; examples include the temperature in a room or the wind velocity in a city. The measurand in an infinite population can take *any* possible value in the given interval. Thus, it is continuously *variable* (not *discrete*) in nature.

In measurement uncertainty, we deal with an infinite population of data/readings of a measurand.

A sampling distribution is the distribution or spread of a sample statistic (usually the mean, but it could be the median, mode, standard deviation, etc.), considered as a random variable, for all possible samples of size n. The shape of the sampling distribution largely depends upon the shape of the population distribution, although owing to the CLT, it usually tends to a Normal distribution for large n (see figure 18.1). It can be seen that as the sample size increases, the sampling distribution

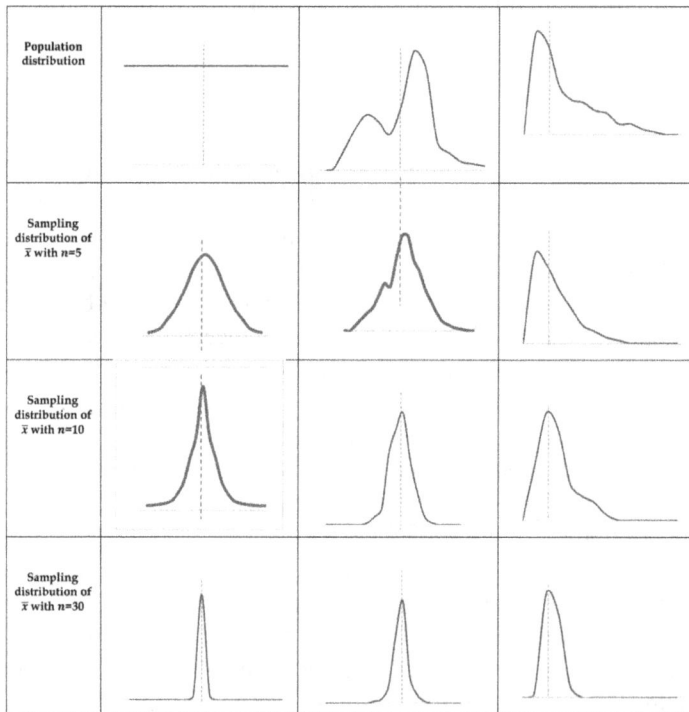

Figure 18.1. Representative sampling distributions of the mean for different sample sizes n.

tends to a Normal distribution, irrespective of the population distribution. (For more on the CLT, see chapter 14, 'How is it that we can combine different distributions?')

A fact one should know:

The statistical variables of a population, such as the mean (μ) and variance σ^2), are called *parameters*, while the corresponding statistical variables of a sample, such as the mean ($\bar{x}$) and variance (s^2) are called *statistics*. Note that the symbols are different.

In general, it may be noted that as the sample size n increases, the distribution of the mean approaches a Normal distribution, irrespective of the distribution of the population. For sample sizes of 30 or more, the distribution of the mean is pretty close to the Normal distribution.

18.1 Sampling error

As we want to estimate the uncertainty of a population parameter based on sampled data, the mean $\bar{x}$ of the sample is bound to be different from the mean of the population μ. This error (difference between the two) arises because $\bar{x}$ is calculated from limited sampled data that has inherent fluctuations or variability. This is called sampling error.

Sampling error arises for the following reasons:
- faulty sample selection
- wrong sampling method
- nonrepresentative sample
- operator bias
- incorrect sample preparation
- heterogeneity or high variability in population

In general, sampling error is inversely proportional to the square root of the sample size (figure 18.2). The larger the sample size, the smaller the sampling error for a given population.

Point estimates and interval estimates:
The value of the statistic (e.g. mean) that we calculate from a sample is a single value and is hence called a point estimate. For every successive sample from the same population, it is most likely that this value will be different for each sample. This error arises due to the dependence on a single sample to draw an inference about the population.

Sampling error depends upon the variability in the population itself and the sample size. Thus, in order to have confidence in whatever single value we have established as a sample mean, we need to have some interval around this value such that this interval may contain the established mean value in a future sample. This is called an interval estimate. Thus, the sample mean has an upper boundary and a lower boundary within which the next sample mean is expected to lie at certain level

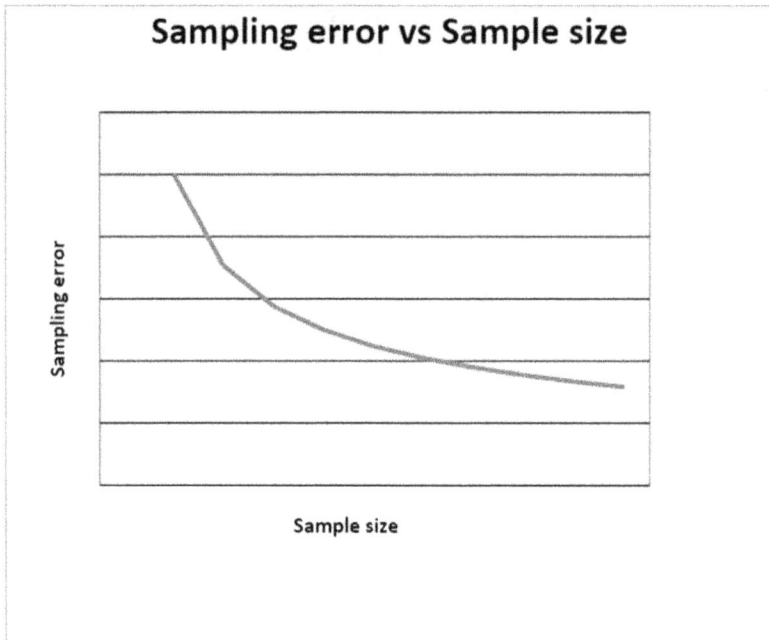

Figure 18.2. Sampling error vs. sample size.

of confidence (usually considered at 95%), assuming that the sample distribution is close to normal.

A fact one should know:

The latest laboratory standard ISO/IEC 17025:2017 [1] has included sampling laboratories in the definition of laboratory. Hence, laboratories that perform only sampling activities can also get accreditation against this standard. However, it covers laboratories that are involved in sampling associated with *subsequent* testing or calibration. In addition to other requirements (clause 7.3 of ISO/IEC 17025:2017), it requires identification of the personnel performing sampling. Hence, the training of the personnel involved in sampling is an important requirement for accreditation. Uncertainty of sampling needs to be considered where appropriate.

Reference

[1] ISO/IEC 17025:2017 *General Requirements for the Competence of Testing and Calibration Laboratories* (Geneva: ISO)

Chapter 19

The sample size dilemma

Abstract: Determining the sample size for repeatability measurements is an issue constantly faced by metrologists. Large amounts of data provide more confidence at the cost of time and resources. Small amounts of data are cost effective but provide little confidence. One needs to strike an optimum balance between the two.

> *There aren't enough small numbers to meet the many demands made of them!*
>
> —Richard Guy

As per the GUM [1], there are two methods for evaluating uncertainty, Type A and Type B. The Type A evaluation of uncertainty is a method that evaluates uncertainty by the statistical analysis of a series of observations. Thus, one needs to conduct an experiment in which one takes a small number of readings under repeatable conditions. The question that usually arises here is: how do we decide how many readings or observations are enough to ensure the credibility of the result?

Let us consider a common experiment to find the average time required for a person to commute between two points in a city. If he/she draws a conclusion from the average of two consecutive days, it may be misleading in the long run (but it is less costly). On the other hand, if he/she takes an average over three months, the conclusion would be more accurate (but the experiment would be more costly). So, what should the length of the experiment be? The answer is to strike a balance between cost and the expected accuracy of the experimental result.

So, for Type A evaluations of uncertainty, the most common questions include: 'What should be the sample size n (number of observations)?' or 'How many readings shall be taken?'

The *ideal* answer would be 'infinite,' but a practical way out is three to five! The first answer is obviously ruled out, which leaves one to wonder whether three to five are enough. Many metrologists often feel that a larger number of readings, say

10–15, would always be better, as this would raise their confidence level. Fundamentally, that is true, as higher and higher numbers of readings would accommodate variations due to extraneous influence factors, leading to a mean value much closer to the 'true' value or the expected value. (By 'true' value, I mean the 'conventional true' value. The 'true' value is a misnomer and it is used here for the sake of understanding. See chapter 6, 'The Guide to the expression of uncertainty in measurement—the new approach' for more on 'true' value as well as clause D 3.5 of the GUM.) Increasing the sample size there is more likelihood of reducing the random error in the measurement.

There is another aspect. If the precision of the measurement is very good (which is albeit subjective and dependent on the application), it does not matter much whether one takes three or 15 readings. But if that is not the case, a larger sample size is always better. But then, how much more is enough? In measurement, we deal with what is called the 'strong law of small numbers,' (formulated by the British mathematician Richard Guy) which states that 'There aren't enough small numbers to meet the many demands made of them!' It means that on one hand, we are compelled to take small numbers of readings due to limitations of resources and business objectives and on the other to extract the maximum information from that data with high confidence! This is a tricky balancing act; however, statistics comes to our rescue. It has been proved statistically that evaluations based on a large number of repeated observations are not necessarily superior to those obtained by other means (i.e. by contributory factors evaluated by the Type B method). This finding is based on the concept of 'relative uncertainty,' which is measured in terms of the relative standard deviation of $s(\bar{q})$ given by the ratio $\{\sigma[s(\bar{q})]/\sigma(\bar{q})\}$. It is approximately given by $[2(n - 1)]^{1/2}$. This 'uncertainty of the uncertainty' arises due to the purely statistical reason of the limited sample size and is surprisingly quite large. For $n = 10$ observations, it is 24%, and for $n = 3$, it is 52%. See table 19.1 (reproduced from the GUM table E.1).

Table 19.1. $\sigma[s(\bar{q})]/\sigma(\bar{q})$, the standard deviation of the experimental standard deviation of the mean $\bar{q}$ of n independent observations of a normally distributed random variable q, relative to the standard deviation of that mean (refer to the GUM for notes). Reproduced from [1] with the permission of the JCGM, which retains full internationally protected copyright.

Number of observations n	$\sigma[s(\bar{q})]/\sigma(\bar{q})$ (%)
2	76
3	52
4	42
5	36
10	24
20	16
30	13
50	10

It is evident from the table that the standard deviation of mean of *observed* samples is pretty high for practical values of sample size n of 3 and 5. Note that even if we increase n to 20, the relative uncertainty is still 16%. Hence, it should be clear that a very large sample size does not certainly add to the reliability of the result nor to our level of confidence in the measurements (the level of confidence *is not* the same as the 'confidence level.' The two terms are not equivalent. See clause 6.2.2 of the GUM). The relative uncertainty for the most common sample size $n = 3$ is a shocking 53%, but due to practical limitations we have to put up with it! At the most, a sample size of five (with a relative uncertainty of 36%) could be a viable option, as it would consume a bit more time without being too taxing in terms of cost. 'A beginner's guide to uncertainty of measurement' [2] by the NPL (UK) recommends four to ten readings. (A higher sample size would certainly reduce the effect of random errors in measurement. See the topics in Part D, 'Degrees of freedom.')

A fact one should know

A further practical consideration related to sample size is the resolution of indication. If the unit under calibration (UUC) resolution is poor, say 3–1/2 digits, or has a least count of one unit (least count is a term indicating lowest count that can be read), and assuming that it has good precision, three readings would suffice, as one would not find any variation. The relative uncertainty discussed above is not of much consequence in such cases. For 4–1/2 digit resolution or higher resolutions, or if the UUC has a count of at least two to three decimal points of a unit, noise and other influence factors would result in variations that would necessitate three to five readings.

Another case in which one can adopt a small sample size is when using a combined or pooled sample standard deviation s_p of a well-characterized measurement system under statistical control. This s_p is the previously estimated standard deviation based on a larger sample size of 10–20 observations and can be used in subsequent measurements (n) taken under essentially similar conditions. The standard uncertainty in such cases is $u = s_p/\sqrt{n}$. However, it should be noted that the scheme of pooled sample standard deviation should be used where comparison measurements are involved or the strict repetition of an observation is not possible due to the different physical locations of the observations, as in hardness testing.

It should be noted that the relative uncertainties in the above table are for an assumed Normal distribution. If this is not the case, the relative uncertainties are even worse!

To sum up, for the Type A evaluation of uncertainty, a sample size of five to ten observations is good enough for most physical measurements.

A fact one should know:

There are some standards that specifically mention the size of the sample, for example, ISO/TS 19036:2006 (Microbiology of food and animal feeding stuffs—Guidelines for the estimation of measurement uncertainty for quantitative determinations) specifies

$n = 10$ as the number of pairs of counts. Another standard, BS 7882:2008 ('Method for Calibration and classification of torque measuring devices'), mentions a specific series/number of steps for observations. Thus, metrologists following such renowned standards must use the sample sizes given by those standards.

Note: ISO/TS 19036:2006 is now replaced by ISO 19036:2019: Microbiology of the food chain—Estimation of measurement uncertainty for quantitative determinations. BS 7882:2008 is now replaced by the 2017 version with the same title [3, 4].

References

[1] JCGM 100:2008 *(GUM) Evaluation of Measurement Data—Guide to the Expression of Uncertainty in Measurement* 1st edn (BIPM, IEC, IFCC, ISO, IUPAC, IUPAP, OIML—International Organization for Standardization)

[2] Bell S 2001 *GPG11 A beginner's guide to uncertainty of measurement* 2nd issue (Teddington: NPL)

[3] ISO 19036:2019 *Microbiology of the Food Chain—Estimation of Measurement Uncertainty for Quantitative Determinations* (Geneva: ISO)

[4] BS7882:2017-TC 2017 *Method for calibration and classification of torque measuring devices* (Chiswick: BSI)

IOP Publishing

A Practical Handbook on Measurement Uncertainty
FAQs and fundamentals for metrologists
Swanand Rishi

Chapter 20

Sample size—another approach

Abstract: Striking a balance between the cost and benefit of a selected sample size is a tricky issue. In this chapter, we take a look at a very simple but effective approach presented by NASA.

> *To call in the statistician after the experiment is done may be no more than asking him to perform a post-mortem examination: he may be able to say what the experiment died of.*
>
> —Ronald Fisher

Sample size is a tricky issue. The mean and standard deviation can vary considerably if the sample size is changed, thus affecting the uncertainty evaluation. Other issues related to uncertainty are normality and outliers (covered separately). In addition, the sample size is important from economic and cost consideration points of view, particularly in the manufacturing industry, where routine checks and measurements are part of quality assurance plans. These may be quite taxing at times. Thus, the sample size has multiple consequences, and selecting an adequate sample size becomes a techno-commercial issue. Small sample sizes are cost effective, but that sample size should also lend credence to the intended technical application. Here, we look at a very simple but effective approach presented by NASA.

The NASA Handbook NASA-HDBK-8739.19-3 [1] provides the following equation to check whether the current sample size, n is sufficient to ensure that the difference between the actual and expected means at given confidence level lies within a certain limit:

$$\sqrt{n} = \left(\frac{\sigma}{c}\right) \quad \varphi^{-1}\left(\frac{1 + \beta}{2}\right) \tag{20.1}$$

As the population standard deviation σ is usually unknown, it is replaced by the sample standard deviation. Thus,

$$\sqrt{n} = \left(\frac{s}{c}\right) \; \varphi^{-1}\left(\frac{1+\beta}{2}\right), \tag{20.2}$$

where
  n = the expected sample size,
  s = the sample standard deviation (of the current measurement),
  c = the sampling error or required difference (or limit) between sample mean
    and population mean,
  φ^{-1} = the inverse normal function and
  β = the confidence level.

The following example illustrates the application of equation (20.2).

Let the temperature readings of a resistance temperature detector (RTD) at 50 °C be as given below:

  50.15 °C, 49.48 °C, 49.90 °C, 50.37 °C, 49.33 °C, 50.60 °C, 50.55 °C, 50.00 °C, and 49.69 °C.

So we have $n = 9$ readings with mean = 50.01 °C and standard deviation $s = 0.45$ °C.

Let us find the minimum sample size required to get a maximum deviation of the sample mean from the population mean of $c = 0.25$ °C at a 95% confidence level. (Although the population mean is unknown, this does not matter.)

Using equation (20.2),

$$\sqrt{n} = \left(\frac{0.45}{0.25}\right) \times \varphi^{-1}\left(\frac{1+0.95}{2}\right)$$
$$= 1.8 \times \varphi^{-1}(0.975)$$
$$= 1.8 \times 1.96$$
$$= 3.53$$

Hence, $n = 3.53^2 = 12.4609$, i.e. a sample size of 13.

Hence, the current sample size of nine is insufficient for the above requirement; we need a minimum of 13 samples. And an increased sample size of 13 will fulfill the requirement, provided the variability (i.e. the standard deviation) remains more or less the same.

An alternate formula that gives the same result as equation (20.2) above is as follows:

$$\sqrt{n} = \left(\frac{s}{c}\right) \times k, \tag{20.3}$$

where k is the coverage factor at a given level of confidence. (The result of solving $\varphi^{-1}\left(\frac{1+\beta}{2}\right)$ is k itself!) In the above example, for a 95% confidence level, $k = 1.96$.

We can take another example. Suppose we want to find the sample size such that the sampling error c is no more than 25% of the standard deviation s at a 99% confidence level.

From equation (20.3), we have

$$\sqrt{n} = \left(\frac{s}{c}\right) \times k.$$

In our example, $c = s \cdot 0.25$ (25% of standard deviation).

Hence,

$$\left(\frac{s}{c}\right) = s/(s \times 0.25) = 4.$$

So,

$$\sqrt{n} = 4 \times (2.58) = 10.32.\ (2.58 \text{ is the coverage factor at a 99\% confidence level.})$$

Thus, $n = (10.32)^2 = 106.5$, i.e. 107.

Points to ponder

Equation (20.2) can be used in a variety of ways to find c, n, or β under various conditions. For example:

(1) The current sample size is sufficient if the requirement for c is reduced to 0.51 °C.

(2) At a 99% confidence level, the sample size would have to be 21.5, i.e. 22.

(3) The current sample size is acceptable at a confidence level of 90%.

If the standard deviation is established using a large sample size, this equation (20.2) can be used to decide upon an economical sample size for given values of c and β. This can be useful in the manufacturing industry, where routine measurements based on sampling have to be made as a part of the QA plan.

Reference

[1] NASA-HDBK-8739.19-3 2010 *Measurement Uncertainty Analysis Principles and Methods, NASA Measurement Quality Assurance Handbook - ANNEX 3* (Washington, DC: NASA)

IOP Publishing

A Practical Handbook on Measurement Uncertainty
FAQs and fundamentals for metrologists
Swanand Rishi

Chapter 21

Addressing the uncertainty of a single measurement

Abstract: In the Type A evaluation of uncertainty, a large number of readings provides higher confidence. However, there are situations in which only a single measurement is possible or called for. This chapter discusses the issues faced in such scenarios.

> *If your experiment needs a statistician, you need a better experiment.*
> —Ernest Rutherford

> *It is better to understand a little than to misunderstand a lot.*
> —George Bernard Shaw

We are generally predisposed to the fact that we need to take a large number of readings (n) to evaluate the repeatability uncertainty. Theoretically, the larger the n, the lower the repeatability uncertainty (which reduces by a factor of $1/\sqrt{n}$). Topics related to sample size are examined in detail in chapter 19, 'The sample size dilemma' and chapter 9, 'Using pooled standard deviation.'

However, there are instances in which we may be required to take two or three readings, or even one reading as an extreme case, in an ongoing measurement. Under any (or a mix) of the following situations, a single measurement may be justified:

1. It is known that the contribution of random errors, including those caused by the measurement device, is negligible.
2. Complexity of measurement.
3. Time constraints.
4. The process or the method employed is well characterized.
5. A large number of devices are to be measured (a routine quality check scenario in manufacturing).

doi:10.1088/978-0-7503-6462-1ch21

6. The measurement of test and measuring equipment (TME) that is used for checking product conformity.

A fact one should know

Situation (6) above is somewhat different from the other situations. TME that is utilized for shop floor checks or for pre-dispatch inspections in the manufacturing industry is used for verifying the conformity of products to specifications. This is mainly a 'go/no go' type of inspection in which the calibration data of the TME (e.g. uncertainty) is of no consequence. Thus, the calibration of such equipment can be done with a single measurement, since the purpose of calibration is to verify that the TME is within its specifications. If the purpose of calibration is to assign a value and estimate uncertainty, a single reading is not the right option.

According to conventional wisdom, if we have only a single measurement, it is difficult to ascertain its reliability. One would be baffled when calculating the standard uncertainty or the degrees of freedom in this scenario. As repeatability uncertainty is evaluated by the Type A method, the formula for its calculation lends itself to this strange situation, as explained below.

The standard deviation of the sample data with n readings is given by

$$s(x_i) = \sqrt{\sum_{i=1}^{n} x_i - \bar{x}^2/(n-1)}. \tag{21.1}$$

For a single reading, $n = 1$. The numerator in this case is zero (as the mean is equal to the single reading itself), and denominator is also zero (as $n - 1 = 0$) in the above equation. And 0/0 is mathematically indeterminate!

What is the solution, then? The problem is circumvented by using past data that is available for similar measurements carried out earlier for a larger sample size m using well-established methods. The GUM [1], which contemplates all practical scenarios, covers this aspect by allowing use of a pool of available information. One such pool of information, as per clause 4.3.1 of the GUM, is 'previous measurement data.' This source of data falls within the Type B evaluation of uncertainty. Clause 4.3.2 of the GUM states that the 'Type B evaluation of standard uncertainty can be as reliable as a Type A evaluation, especially in a measurement situation where a Type A evaluation is based on a comparatively *small* number of statistically independent observations.' However, it requires good knowledge of the method, use of experience, and proper judgment.

Thus, when $n = 1$ in the current measurement, the uncertainty obtained from a previous similar method with m readings is used for Type A (i.e. repeatability) evaluation. The standard deviation of the previously estimated sample data based on m measurements is

$$s(x_i) = \sqrt{\sum_{i=1}^{m} x_i - \bar{x}^2/(m-1)}. \tag{21.2}$$

(This standard deviation is readily available for current use.)

The experimental standard deviation of the mean in the current set of n measurements is given by $s(\bar{x}) = s(x_i)/\sqrt{n}$, where $s(x_i)$ is taken from equation (21.1).

A fact one should know

Obviously, for a single reading, $n = 1$, and the standard uncertainty $s(\bar{x})$ for a Type A evaluation is same as $s(x_i)$. Note that the degrees of freedom in such cases are $m - 1$ (and *not* $n - 1$, which would be ridiculous for a single measurement!), where m is the number of measurements in the *prior* evaluation. Although we take n readings in the current measurement, the degrees of freedom are $m - 1$ because we utilize the uncertainty obtained from m measurements of previous data. (Also see Part D, 'Decoding degrees of freedom' for other details.)

Let us take an example of the measurement of rod diameter on a production floor in a company. The operator takes a single reading of the rod diameter selected at predetermined intervals on the conveyor belt. Let that reading be 27.65 mm. If an earlier estimate of a similar experiment repeated with $m = 10$ readings gave an uncertainty of 0.12 mm, the uncertainty of the current reading is also 0.12 mm. Hence, the result is given as:

rod diameter = 27.65 mm ± 0.12 mm with degrees of freedom = 9.

Such methods are well characterized for m measurements (m is usually large, say $10 - 15$), since the data obtained from them are (also) meant for future use in routine measurements of a similar type. A large number of readings in a prior evaluation gives a more reliable estimate when only a few (or only one, in the current discussion) measurements can be made during the routine procedure.

The reliability of the previous data depends on the number of devices sampled, the number of readings taken, and how well this sample represents all devices. Such prior data should be regularly reviewed. A previous estimate of standard deviation can only be used if there has been no subsequent change in the measurement system or procedure that could affect repeatability. Hence, the regular validation of such a procedure has to be ensured.

A fact one should know

Clauses F.2.4.1 and F.2.4.2 of the GUM hint at the validity of a single observation against a calibrated instrument or a verified instrument respectively, under certain circumstances. Clause 4.11 of UKAS M3003:2007 [2] suggests taking at least two measurements but also accepts a single measurement, even with imperfect repeatability. It recommends that one should rely on previous assessment for similar devices, provided that the precautions explained earlier are taken. It also cautions that if a large spread is observed, the cause should be investigated and resolved before proceeding further. (This may create confusion, as 'spread' would be a misnomer in the case of a single observation. One should instead look for a 'deviation' from the expected reading.)

21.1 How do we proceed if no such information of previous data or history is available?

As explained in the opening part of this topic, a few situations are typically found in manufacturing activity on the shop floor. In addition to the first situation, there is a time constraint. Further, there could be a scenario in which such past history or information is not available.

For example, referring to the earlier rod diameter example, let us assume that a regular batch has a nominal diameter of, say, 27.60 mm. There is an urgent new order for a rod diameter of 25.40 mm in a limited quantity. In such cases, the past information for a batch that had a rod diameter of 27.60 mm (or any other similar diameter) may be used. This requires confirmation that the current production process capability for a rod diameter of 25.40 mm, the environmental conditions, the material properties, and the client's uncertainty requirements are similar or close to those of the 27.60 mm facility. In the absence of this confirmation, a well-characterized evaluation under controlled conditions with a large (> 10) number of readings is required before the order can be accepted. This data could be used in further orders for same rod diameter with only a single measurement.

References

[1] JCGM 100:2008 *(GUM) Evaluation of Measurement Data—Guide to the Expression of Uncertainty in Measurement* 1st edn (BIPM, IEC, IFCC, ISO, IUPAC, IUPAP, OIML—International Organization for Standardization)
[2] UKAS M3003 2007 *The expression of uncertainty and confidence in measurement* (Staines-upon-Thames: UKAS) [The latest edition is of March 2024]

Part D

Decoding degrees of freedom

Chapter 22

Why is degrees of freedom generally $(n - 1)$ in type A method of evaluation?

Abstract: The degrees of freedom is a measure of uncertainty of the variance. A high number of degrees of freedom is associated either with a large number of repeat measurements (Type A evaluation) or very high level of confidence (practically 100%) on the assumed probability distribution (Type B evaluation). The degrees of freedom ν are $n - 1$ for a *single* quantity estimated by taking the arithmetic mean of n independent observations. For a least-squares fit of m parameters to n data points, the degrees of freedom of the standard uncertainty of each parameter are $\nu = n - m$.

> *I can prove anything by statistics—except the truth.*
>
> —George Canning

When estimating measurement uncertainty, it is necessary to mention the degrees of freedom of each component. As noticed during seminars, it is one of the most difficult terms to comprehend. (*Caution! It is 'degrees,' always plural.*) It is customary to take the degrees of freedom for the Type A method of evaluation as equal to $(n - 1)$, where n is the sample size. Thus, for a sample size of $n = 10$, the degrees of freedom are 9. This is done ritually; however, this is valid in most of the routine measurements.

The degrees of freedom ν is equal to $n - 1$ for a *single* quantity estimated by the arithmetic mean of n independent observations. If n independent observations are used to determine two quantities, say by the least-squares method, the degrees of freedom of their respective standard uncertainties are $\nu = n - 2$. In general, for a least-squares fit of m parameters to n data points, the degrees of freedom of the standard uncertainty of each parameter are $\nu = n - m$.

However, why are the degrees of freedom given by $n - 1$ and not $n - m$, where m could be any other number $< n$? This concept is fundamental in measurement uncertainty and has great significance for the results estimated.

By definition (ISO 3534-1, 2.85) [1], the degrees of freedom is given by 'the number of terms in a sum minus the number of constraints on the terms in the sum.' The 'number of constraints' means the number of *statistics* we estimate using that data. In Type A evaluation, we start from n data points. From this data, we find the mean and then proceed to estimate the standard deviation—the intended statistic. Thus, in evaluating standard deviation, we have $(n - 1)$ degrees of freedom.

The standard deviation of the sample data is given by

$$s = \sqrt{\sum_{i=1}^{n}(x_i - \bar{x})^2/n - 1}. \tag{22.1}$$

The term $(n - 1)$ appears in the denominator and represents degrees of freedom. It arises from the correlation between x_i and $\bar{x}$ and signifies the fact that there are only $(n - 1)$ independent items in the set of n data points.

The degrees of freedom appearing in a t-distribution is a measure of the uncertainty of the estimated variance and is in inverse proportion to it. In other words, the greater the degrees of freedom, the less the variance.

A fact one should know

The degrees of freedom is a measure of the uncertainty of the variance, and the latter is inversely proportional to the former (refer to equation (22.1)). In measurements, we generally collect data to establish one or two quantities. For example, to evaluate uncertainty in measurement, we establish only the experimental standard deviation of the mean (a single quantity) from the mean of n independent data points. However, if the data is utilized to establish one more quantity, the degrees of freedom are $(n - 2)$! A common example is the use of the least-squares method to establish both the slope and the intercept of a line. In general, if we establish m parameters from n data points, the degrees of freedom of the standard uncertainty of each parameter $= (n - m)$. The number of constraints in the definition of degrees of freedom are the number of parameters m that we obtain from n data points.

The degrees of freedom can also be explained in another way. If we have five data points, say 1, 2, 3, 4, and 5, their mean is 3. Once we know the *mean*, using *any* four data points, we can find the fifth point. Therefore, there are four (i.e. $(n - 1)$) *independent* points and the fifth can be established automatically. Therefore, the degrees of freedom indicate the number of independent comparisons one can do with the sample data. Although there are n independent samples, there are only $(n - 1)$ independent *residuals.*

The degrees of freedom is also a measure of the 'freedom' that one gets while maneuvering or handling the data. Naturally, the more the data, the more the freedom. It also quantifies the amount of information available to us. When the degrees of freedom is large, it means we have estimated the result from a large amount of data. It is obvious that as n increases, $n \approx (n - 1)$ and $(n - 1)$ being in the denominator in equation (22.1), the standard deviation of the mean approaches the standard deviation of the population, i.e. the difference between them narrows down.

A large number of degrees of freedom is an insignia of a higher confidence level for the collected data. In classical statistics, it is a measure of the reliability of the sample standard deviation as an estimate of the population standard deviation. The GUM adopted this concept to quantify u_c as a representation of the standard deviation of the error distribution. We undoubtedly have more faith in 20 data points than in five or ten (or even 19 for that matter). A large number of degrees of freedom in a Type A evaluation also helps us to obtain a large number of effective degrees of freedom, typically > 30, so that we can safely assume a Normal distribution rather than a t-distribution. A high number of degrees of freedom is associated either with a large number of measurements or a value associated with a low variance or low dispersion.

A fact one should know

For a Normal distribution, the sample mean and the sample variance are *unbiased* estimators of the population mean μ and the population variance σ^2, respectively. (An estimator is unbiased if it is equal to its expectation, E, or has no bias. An unbiased estimator is deemed to be a *good* estimator.) The sample variance s^2 turns out to be a biased estimator if n is used in the denominator, but it can be corrected (made unbiased) by using $n - 1$. This is another reason that we use $(n - 1)$ and not n in the denominator to calculate the sample variance. However, the sample standard deviation s is not an unbiased estimator, as it is a square root of variance, and square root is *concave downwards*. Thus, s as an estimator of σ is downwardly biased. Obviously, the extent of the bias decreases with increasing sample size.

When a pooled standard deviation based on N series of observations is used for calculations, the degrees of freedom are given by

$$\nu = \sqrt{\sum_{i=1}^{N} \nu_i},$$

(22.2)

where ν_i are the degrees of freedom of the ith series of n_i independent repeat observations.

Reference

[1] ISO 3534-1 2006 *Statistics—Vocabulary and Symbols—Part 1: General Statistical Terms and Terms Used in Probability* (Geneva: ISO) [Confirmed in 2021]

Chapter 23

Why is degrees of freedom generally '∞' in type B evaluation?

Abstract: In most practical applications, the degrees of freedom for the Type B evaluation of standard uncertainty is assumed to be infinite. This indicates complete confidence in the data of the uncertainty component considered for the evaluation of uncertainty. This chapter discusses the concepts behind this.

> *Everybody is a Bayesian. It's just that some know it, and some don't.*
> —Trivellore Raghunathan

As explained in an earlier topic, the degrees of freedom indicates a confidence level on the information used for uncertainty evaluation. Unlike Type A evaluation, which is *frequency based*, we do not use 'sample' data in Type B evaluations; rather, we rely on all other available data and information. Traditional statisticians did not treat a Type B estimate as a statistic in the 'pure' sense. Hence, it was not considered to have definable degrees of freedom that could be regarded as quantifying the amount of information used for the evaluation. Accordingly, if used alone or combined with a Type A estimate, the result was not viewed as being a 'true' statistic. However, as shown in appendix E of the GUM [1], Type B estimates can be as reliable and valid as Type A estimates, because the latter have a large 'uncertainty of uncertainty' if the sample size is small. (See table E.1 of the GUM. This is also covered in chapter 19, 'The sample size dilemma.')

In Type B evaluation, where we generally consider the degrees of freedom to be ∞, it means our level of confidence is extremely high. Rather, we have 100% confidence in the data or the assumption of probability distribution. For example, when we assume a uniform distribution which is characterized by lower and upper boundary, we strongly believe that the probability that the data lie within the bounds is 100%. In Type B evaluation, we also refer to the uncertainty of the reference standard from its certificate and assign the degrees of freedom as ∞. This is because the laboratory arrived at the results with effective degrees of freedom more than 30, giving a coverage factor very close to the value obtained at degrees of freedom equal to ∞. (However, clause G.6.6 of the GUM mentions that when the effective degrees of freedom are more than ten, the uncertainty of the estimated combined uncertainty $u_c(y)$ is reasonably small.) The difference and the consequence of taking the exact coverage factor become less significant for effective degrees of freedom more than 30. (See chapter 24, 'Effective degrees of freedom—some considerations.')

A fact one should know

It is customary to assume that the degrees of freedom for a Type B component are ∞. However, it should be noted that in Type B evaluation, it is possible to comprehend degrees of freedom other than ∞! For this, let us make use of equation G.3 of the GUM, which defines degrees of freedom as

$$\nu_i = \frac{1}{2} \left[\frac{\Delta u(x_i)}{u(x_i)} \right]^{-2}.$$

The quantity in square brackets represents the *relative* uncertainty of $u(x_i)$. In Type B evaluation, we evaluate the standard uncertainty components by considering them to be *a priori* probability distributions and greatly rely on scientific judgment and *degree of belief*. This degree of belief, which we assign for the probability distributions or the data, need not always be 100%. For example, one may believe that the temperature coefficient of an uncommon metal mentioned in a published paper could be reliable to, say, 20%, which could be considered as its relative uncertainty. Thus, while estimating its contribution using the above equation, the degrees of freedom are $\frac{1}{2}[0.20]^{-2}$, i.e. 50; alternatively, if it is reliable to 50%, the degrees of freedom are 2. Thus, in Type B evaluation, we obtain degrees of freedom from the judgment of reliability of the value of that component. Hence, it should be realized that when we take the degrees of freedom to be ∞ in Type B evaluation, we implicitly assume that value of $u(x_i)$ is *exactly* known. Of course, this is nevertheless deemed to be quite realistic. The latter case (degrees of freedom $= \infty$) is safer (or desirable—being convenient!) than using the above equation, since it results in higher effective degrees of freedom as per the Welch–Satterthwaite formula. Certainly, one needs to be judicious, keeping in mind that scientific judgment should prevail over the out-and-out degree of belief!

Reference

[1] JCGM 100:2008 *(GUM) Evaluation of Measurement Data—Guide to the Expression of Uncertainty in Measurement* 1st edn (BIPM, IEC, IFCC, ISO, IUPAC, IUPAP, OIML—International Organization for Standardization)

IOP Publishing

A Practical Handbook on Measurement Uncertainty
FAQs and fundamentals for metrologists
Swanand Rishi

Chapter 24

Effective degrees of freedom—some considerations

Abstract: The degrees of freedom of the Type A and Type B methods of evaluation of uncertainty are *usually* $n - 1$ and ∞, respectively. When the uncertainties obtained by two methods are combined, the resultant degrees of freedom are called 'effective degrees of freedom.' The chapter deliberates on this concept and its practical considerations.

> *He uses statistics as a drunken man uses lampposts—for support rather than for illumination.*
>
> —Andrew Lang

For practical purposes, we take three to five sample readings for the Type A evaluation of uncertainty. This precludes us from reaching a high confidence level, as the degrees of freedom is quite small. We further take recourse to Type B data, as we believe that apart from the observed variance, other known factors affect our results. A limited number of readings also precludes us from using a strictly Normal distribution. Rather, in such cases, a *t-distribution* is the right choice, which approaches a Normal distribution, with mean, $\mu = 0$ and standard deviation, $\sigma = 1$ as ν_{eff} tends to ∞.

The Working Group of the CIPM not only recommends that combined uncertainty, u_c should be used for the result of measurements but also recommends that it should be used for all international comparisons and works under its auspices. Appendix C of the NIST TN 1297 [1] policy also states that u_c should be used to report the results of determinations of fundamental constants, fundamental metrological research, and international comparisons of realizations of SI units.

However, clause 6.1.2 of the GUM [2] takes cognizance of the practical aspects of using uncertainty in trade, industry, health, safety, regulatory, and commercial applications, in which it is necessary that uncertainty should encompass 'a large fraction' of the distribution of values around the measurand. This requirement is met

doi:10.1088/978-0-7503-6462-1ch24

through the use of 'expanded uncertainty.' This is done to accommodate variations observed in the above areas due to practical limitations as well as to account for any unknown influence factors left out during evaluation.However, it should not be understood as assigning a safety limit so as to enlarge combined uncertainty.

For a normally distributed variable z, the distribution of the variable $t = (\bar{z} - \mu_z)/s(\bar{z})$ is a *t-distribution* (named after William Gosset, who took the nickname 'Student'; therefore the *t*-distribution is also called Student's distribution) with $\nu = (n - 1)$ degrees of freedom. Due to limited sample data and the availability of only the *estimates* of quantities (instead of the quantities themselves), we use an approximation for the distribution of the variable Y. Hence, instead of a distribution of $[Y - E(Y)]/\sigma(Y)$, we use the distribution $(y - Y)/u_c(y)$. The problem arises because this does not describe the distribution of the variable $(y - Y)/u_c(y)$ if $u_c^2(y)$ is sum of two or more estimated variances of input components, even if they are normally distributed. When Y (the measurand) is a single normally distributed quantity X, this distribution reduces to the distribution given by $(y - Y)/u_c(y)$. This can be obtained, to the lowest-order approximation, using the Welch–Satterthwaite formula.

However, according to the central limit theorem, (CLT), the distribution $(y - Y)/u_c(y)$ may be approximated by a *t-distribution* if:

- the distribution of Y is normal,
- both u_c and y are independent, and
- the distribution of u_c^2 is a χ^2 distribution.

As stated in the GUM, first two conditions can be approximately assumed to be satisfied in most practical scenarios. The last condition is satisfied by expanding $(y - Y)/u_c(y)$ in a Taylor series about its expectation with *effective* degrees of freedom ν_{eff}.

The Welch–Satterthwaite formula for effective degrees of freedom, is given by

$$\nu_{\text{eff}} = \frac{u_c^4(y)}{\displaystyle\sum_{i=1}^{n}(u_i^4(y)/\nu_i)}. \tag{24.1}$$

where, u_i and ν_i are standard uncertainty and degrees of freedom of ith component of uncertainty respectively. Usually, ν_{eff} is not an integer, in which case it is either interpolated or truncated to the next lower integer.

Facts one should know

FACT 1: Although the Welch–Satterthwaite formula is applicable for statistically independent, normally distributed error sources, it is a reasonable approximation even when error sources are not statistically independent. In the Welch–Satterthwaite relation, u_c is computed by assuming that no error correlations exist, all contributions are approximately equal in size, and that the sensitivity coefficients for the measurement process uncertainties are all equal to one.

FACT 2: One may ask why we need to calculate effective degrees of freedom. Why not just take them to be ∞? The answer is that if we take the degrees of freedom, $\mathrm{DoF} = \infty$ irrespective of their actual value, we end up with a coverage probability that is *less* than the actual probability. This effect is more prominent when the sample size is small. From Student's t-table, we see that the t-factor increases rapidly as the DoF become smaller, implying that we need to contemplate a higher coverage probability. See FACT 2 in the next box below.

Also note that DoF are required when the uncertainty is estimated as per the GUM (JCGM 100) [2]. Where another approach, such as the Monte Carlo Method of JCGM 101 [3] is employed, DoF are not required, as that method is based on the propagation of the *distributions* themselves instead of the distributions of *estimates* used in the GUM (JCGM 100). See chapter 34, 'Alternative approaches in uncertainty evaluation.'

The expanded uncertainty is obtained by multiplying the combined uncertainty by a coverage factor k that is based on: (a) the level of confidence at which we want to specify the expanded uncertainty and (b) the effective degrees of freedom, ν_{eff}. This is achieved using Student's t-statistic.

Note that even if the expanded uncertainty is not required to be given, it is mandatory to state ν_{eff}, since the GUM requires that degrees of freedom shall be mentioned for every uncertainty component, including u_{c}.

Facts one should know

FACT 1: ISO 21748:2017 [4] gives helpful guidance on deciding the effective degrees of freedom: 'Where any one term is dominant (i.e. when it is roughly > 0.7 times $u_{\mathrm{c}}(y)$) it is normally sufficient to consider effective degrees of freedom for $u_{\mathrm{c}}(y)$ equal to that of the dominant term.'

FACT 2: Clause G.6.6 of the GUM mentions that in many practical measurements in a broad range of fields, when the effective degrees of freedom is more than ten, the uncertainty of the estimated combined uncertainty $u_{\mathrm{c}}(y)$ is *reasonably small*. This means that more than ten degrees of freedom are significantly large, and one can deduce that they could be treated as equal to ∞. Thus, one can use $k = 2$ to obtain the expanded uncertainty at a level of confidence of approximately 95%. However, at a 95% level of confidence, for $\nu_{\mathrm{eff}} = 11$, the deviation of t-factor from the t-factor at ∞ is on the high side by about 10% (and by about 12.5% at a 95.45% level of confidence). Thus, taking $k = 2$ for $\nu_{\mathrm{eff}} = 11$ causes under-evaluation in some applications and may not be acceptable.

As an addendum, the GUM clarifies that: 'in practice, the size of ν_{eff} and what is required of expanded uncertainty will determine whether this approach can be used.'

In general, it is proposed (or is better) to take effective degrees of freedom $\geqslant 30$ as approximating ∞. For these degrees of freedom, the coverage factor $k = 2.09$ at 95.45% level of confidence deviates by just 4.5% of the t-value at degrees of freedom equal to ∞. An overevaluation of about 5% is reasonably small and should be acceptable in most practical situations. (Incidentally, for effective degrees of freedom $= 30$, the arithmetic addition of the deviation (4.5%) to the corresponding level of confidence (95.45%) gives 99.95%, which is as good as 100%! Thus, at $\nu_{\mathrm{eff}} \geqslant 30$, our 'degree of belief' can be 100%!)

References

[1] NIST 1994 *Guidelines for Evaluating and Expressing the Uncertainty of NIST Measurement Results* (Gaithersburg, MD: NIST) Technical Note 1297

[2] JCGM 100:2008 *(GUM) Evaluation of Measurement Data—Guide to the Expression of Uncertainty in Measurement* 1st edn (BIPM, IEC, IFCC, ISO, IUPAC, IUPAP, OIML—International Organization for Standardization)

[3] JCGM 101:2008 *Evaluation of Measurement Data—Supplement 1 to the 'Guide to the Expression of Uncertainty in Measurement'—Propagation of Distributions Using a Monte Carlo Method* (Sèvres: BIPM)

[4] ISO 21748:2017 (confirmed in 2022) *Guidance for the use of repeatability, reproducibility and trueness estimates in measurement uncertainty evaluation* (Geneva: ISO)

Part E

Some contiguous concepts

IOP Publishing

A Practical Handbook on Measurement Uncertainty
FAQs and fundamentals for metrologists
Swanand Rishi

Chapter 25

What is the significance of the sensitivity coefficient?

Abstract: The sensitivity coefficient describes how the output estimate y varies with changes in the values of the corresponding input estimate x_i when the latter is changed by a small amount Δx_i. It can be calculated by taking the partial derivative of the function for the output quantity y with respect to the input quantity x, $\left(\frac{\partial f}{\partial x_i}\right)$ or by doing practical experiments in which $\frac{\Delta y}{\Delta x_i}$ is the sensitivity coefficient.

> *Do not put your faith in what statistics say until you have carefully considered what they do not say.*
>
> —William W Watt

When evaluating measurement uncertainty, we mention the sensitivity coefficient in the uncertainty budget. We multiply the standard uncertainty of the input estimate $u(x_i)$ by the sensitivity coefficient (c_i) to get the uncertainty contribution $u_i(y)$. Most of the time, with a few exceptions, the sensitivity coefficient for Type B evaluations is close (or equal) to one. Since any number multiplied by one is the same number, why is the mention of sensitivity co-efficient required at all? To understand, let us examine some theory.

Individual uncertainty contributions are combined using the *law of propagation of uncertainty*. This states that the uncertainties of the input quantities taken equal to the standard deviations of their probability distributions combine to give the uncertainty of the output quantity if that uncertainty is taken equal to the standard deviation of the probability distribution of the output quantity. The combined uncertainty (for uncorrelated input quantities, a common case) is given by

$$u_c^2(y) = \sum_{i=1}^{N} \left(\frac{\partial f}{\partial x_i}\right)^2 u^2(x_i). \qquad (25.1)$$

doi:10.1088/978-0-7503-6462-1ch25

The partial derivatives $\frac{\partial f}{\partial x_i}$ are called sensitivity coefficients and describe how the output estimate y varies with changes in the input estimates ('∂' stands for partial derivative). Equation (25.1) can be given in a simplified form as

$$u_c^2(y) = \sum_{i=1}^{N}[c_i u(x_i)]^2, \tag{25.2}$$

where $c_i = \frac{\partial f}{\partial x_i}$ stands for the sensitivity coefficient. This requires the individual variance terms in the propagation of error formula to be *multiplied* by the sensitivity coefficients c_i. (It should be noted that equation (25.2) is based on a first-order Taylor series approximation of the measurand Y. When the nonlinearity is significant, higher-order terms must be included in that equation.)

The measurement uncertainty evaluation method given in the GUM [3] is generic in nature. i.e. its methodology takes into account all possible scenarios in measurement. Hence, although the sensitivity coefficient is usually one, there are some exceptions.

The relation between the input and output estimates can be of three types.

1. The input and output quantities are the same. For example, when calibrating a temperature indicator that gets input from a PT100 sensor, we can give the direct temperature input (in units of, say, °C) from the calibrator, and the indicator also displays the temperature in °C (the units are the same).
2. The input and output quantities are different due to the specific measurement method. For example, when calibrating the same temperature indicator, we can give the resistance input (in units of ohms) from the calibrator, but the indicator displays the temperature in °C (the units are not the same).
3. The input and output quantities are different due to the specific nature of the device under test (DUT). For example, while calibrating a temperature transducer, we can give the temperature input (in units of °C) from the calibrator and measure the output of the transducer in units of milliamperes (again, the units are not the same).

(The first two cases are for a device under calibration (DUC) in 'measure' mode but also apply to 'Source' mode.)

In the second and third cases, the sensitivity coefficient is usually not equal to one because there is specific conversion factor. For example, for a temperature transducer that has a 4–20 mA output for a given input temperature range, the conversion factor is given in the form of 'x' mA/°C. Thus, in such cases, the sensitivity coefficient also helps to harmonize the units of measurements of the input and output quantities.

The sensitivity coefficient is also defined as the differential change (Δy) in the output estimate that is generated by a differential change in an input estimate divided by the change (Δx) in that input estimate. So, the sensitivity coefficient $= \frac{\Delta y}{\Delta x}$. Thus, it is also possible (see clause 5.4.1 of the GUM) to find the sensitivity coefficient experimentally by measuring the change in output produced by a small change in a particular input while holding the remaining input quantities constant. This activity is specifically appropriate when the functional relationship between the input and the output is complex and the calculation of partial derivatives is thus complicated.

Facts one should know

FACT 1: When empirically determining the sensitivity coefficient, a small change in the input (Δx) may sometimes not produce a detectable change (Δy) in the output. This could happen because of poor resolution of indication or the equipment may have poor sensitivity or linearity. For example, for an indicator that has a resolution of 0.1 mV, a 'small' change of 10 μV in the input does not have any effect on the output indication. (But this does not mean that the sensitivity coefficient $= 0$.) The change may be detectable or 'visible' at, say, 98.6 μV, resulting in a sensitivity coefficient of 0.1 mV/ 0.0986 mV $= 1.01 \approx 1$. So, the 'small' change in input should be 'big' enough to get a meaningful sensitivity coefficient! Therefore, the resolution does matter when we determine the sensitivity coefficient by experiment. What if it takes a 50 μV input to produce a 0.1 mV change in the output? The sensitivity coefficient is two! Should we use this in calculations? Surely not, since we would not be required to reach that stage of evaluation. The reason for this is that before we began the sensitivity coefficient business, the problem would have been detected in initial checks by getting readings that were substantially out of specification, by almost 200%.

FACT 2: When the sensitivity coefficient is established experimentally, the GUM (per clause 7.2.7) advises that this fact should be revealed when reporting the results.

Also see second box— A fact one should know in chapter 33, 'The proper reporting of uncertainty,' which discusses a different aspect of the sensitivity coefficient.

To summarize, in most practical cases (as in the example in FACT 1), the functional relation between the input and output estimates is linear over a short range if the units of the input and output estimates are the same. Hence, the sensitivity coefficient is usually one.

It is $\neq 1$ when:
- (a) the units of the input influence quantity and the output estimate are not the same (e.g. a transducer), or
- (b) the functional relation between the input and output estimates is nonlinear (e.g. $P = V^2/R$), or
- (c) a comparison measurement is done (e.g. comparison with a CRM), or
- (d) the output quantity is derived using a mathematical relationship between the output and measured input estimates.

An example for the evaluation of uncertainty for a Brinell Hardness Tester illustrates the last point (d) above. It shows how sensitivity coefficients are useful for the conversion of different units to the same unit.

The equation for Brinell hardness is given by

$$B = 0.204 \times F / \left[\pi \times D \left\{ D - \sqrt{(D^2 - d^2)} \right\} \right] \text{N mm}^{-2} \text{ (Note the unit).}$$

There are three factors that contribute to uncertainty:
1. Force, F, with a **unit of N**.
2. Ball diameter, D, with a **unit of mm**.

3. Indentation diameter, d, with a **unit of mm**.

The units of the uncertainties of the respective factors will also be the same. However, we should get uncertainty in the unit of B, i.e. in N mm^{-2}.

The combined uncertainty for B is given by

$$u_B = \sqrt{\{U_F^2 + U_D^2 + U_d^2\}} \text{ N mm}^{-2},$$

where U_F, U_D, and U_d are the respective uncertainty contributions.

- The sensitivity coefficient for force is

$$c_F = \partial B/\partial F \text{ ('}\partial\text{' stands for } partial \text{ derivative)}$$
$$= 0.204/[\pi \times D\{D - \sqrt{(D^2 - d^2)}\}] \text{ mm}^{-2}.$$

Hence, the uncertainty contribution due to force

$$= c_F = u_F \times c_F \text{ N mm}^{-2}.$$

- The sensitivity coefficient for the ball diameter is

$c_D = \partial B/\partial D$
$$= 0.204 \times F\{[D^2/\sqrt{(D^2 - d^2)}] + [\sqrt{(D^2 - d^2)} - 2D]\}/[\pi \times D^2\{D - \sqrt{(D^2 - d^2)}\}^2] \text{ N mm}^{-3}.$$

Hence, the uncertainty contribution due to the ball diameter

$$= U_D = u_D \times c_D \text{ N mm}^{-2}.$$

- The sensitivity coefficient for the indentation diameter is

$c_d = \partial B/\partial d$
$$= -0.204 \times F \times d/\{\pi * D\sqrt{(D^2 - d^2)} \times [D - \sqrt{(D^2 - d^2)}]^2\} \text{ N mm}^{-3}.$$

Hence, the uncertainty contribution due to indentation

$$= U_d = u_d \times c_d \text{ N mm}^{-2}.$$

Thus, the different units of the contributory factors ('newtons' for force and 'mm' for diameter) have been converted to same unit as that of Brinell hardness (N mm^{-2}) using sensitivity coefficients. Hence, the result and its uncertainty can be given in the same unit.

References and further reading

[1] NIST 1994 *Guidelines for Evaluating and Expressing the Uncertainty of NIST Measurement Results* (Gaithersburg, MD: NIST) Technical Note 1297
[2] ISO 21748 2017 *Guidance for the Use of Repeatability, Reproducibility and Trueness Estimates in Measurement Uncertainty Evaluation* (Geneva: ISO)
[3] JCGM 100:2008 *(GUM) Evaluation of Measurement Data—Guide to the Expression of Uncertainty in Measurement* 1st edn (BIPM, IEC, IFCC, ISO, IUPAC, IUPAP, OIML— International Organization for Standardization)

IOP Publishing

A Practical Handbook on Measurement Uncertainty
FAQs and fundamentals for metrologists
Swanand Rishi

Chapter 26

Dealing with corrections

Abstract: Corrections for known systematic effects have to be applied in measurement. In spite of corrections, what we get is the best estimate of the true value. The GUM [1] repeatedly insists upon the application of corrections. However, after the application of corrections, even if the error is reduced to zero, the uncertainty of the correction remains and needs to be accounted for. A zero error does not mean zero uncertainty.

> *To be sure of hitting the target, shoot first and call whatever you hit the target.*
>
> —Ashleigh Brilliant

There had been practice in the past to neglect small corrections in the result and reflect them in uncertainty. This correction was considered to be a rectangular distribution. This poses difficulty in using the result as well as uncertainty in further measurements. Sometimes, a few small corrections can result in a sizable error in the result and may be comparable to the uncertainty.

A correction (GUM B.2.23) is a 'value added algebraically to the uncorrected result of a measurement to compensate for systematic error.' The VIM [2] defines it in clause 2.53 as 'compensation for an estimated systematic effect.' So the corrections can be applied for *systematic* effects only.

The GUM has repeatedly advocated the application of correction (or a correction factor; see GUM B.2.24) to measurement results impacted by systematic effects. (However, it may be noted that the corrections cannot be applied for random effects.) Clause 3.2.4 says that 'It is assumed that the result of a measurement has been corrected for all recognized significant systematic effects and that every effort has been made to identify such effects.' Under practical conditions, the result of measurement (the mean value obtained with a limited sample size) is supposed to be the *best* estimate of the measurand. To support this assumption and provide assurance, it is imperative that this result 'truly' represents the measurand. By correcting known systematic effects, we try to bring this value into agreement with its 'true' value and thus reduce the bias. Without the correction of systematic effects, the result will have an unwarranted deviation from the reference value. In addition, it will not be reliable and useful for further treatment. Correction needs to be applied because the GUM assumes that after it is applied, the expectation or expected value of the error arising from systematic effects is zero. However, even if the error is reduced to zero, the uncertainty of correction remains and needs to be accounted for. A zero error does not mean zero uncertainty.

In clauses 6.3.1 and F.2.4.5, the GUM discourages the practice of showing the correction separately in the final result, unless the expenses of applying correction are unacceptable. The GUM clarifies that a separate correction can be shown as a replacement of an expanded uncertainty U with $U + b$, (where U is an expanded uncertainty obtained under the assumption $b = 0$), only where *all* of the following conditions apply:

1. the measurand Y is defined over a range of values of a parameter t, as in the case of a calibration curve for a temperature sensor;
2. the uncertainty 'U' and the correction 'b' also depend on t; and
3. only a single value of 'uncertainty' is to be given for all estimates $y(t)$ of the measurand over the range of possible values of t.'

In such a situation (as an exception and where all the above conditions are satisfied) the GUM offers the simple approach of applying a *single* correction b to all values of parameter t, giving the final result $Y(t) = y'(t) \pm U = y(t) + \bar{b} \pm U$, where $y(t)$ is the uncorrected estimate of $Y(t)$ and $\bar{b}$ is a single mean estimated by integrating $b(t)$ over the range from t_1 to t_2. However, in such cases, it advises that it should be clearly stated that 'a *single average correction* has been applied over entire range.'

EUROLAB's Technical Report No. 1/2007 [3] agrees with the GUM's perspective in clause F.2.4.5 on correction, but takes refuge in the words 'known corrections' to differ on its implementation. It contends that 'In practice, however, it will often be the expenses for *deriving* rather than for *applying* a 'known correction' that are prohibitive. Then increasing measurement uncertainty to account for significant bias is most certainly better than applying a doubtful correction or, even worse, ignoring the bias.' Example H.3 in the GUM presents a method for applying corrections to each value in the range rather than making a single correction (which is a simple alternative as per F.2.4.5) to the entire range. The exhaustive and tedious treatment vindicates the apprehension mentioned in EUROLAB's report cited above. However, such a treatment could be justified for critical applications.

A simple example will illustrate how to apply correction. The length of a gauge block is to be measured using a micrometer. Let:

x_n = the nominal length of the gauge block = 20 mm
α_g = the thermal coefficient of linear expansion of the gauge block = $11.5 \times 10^{-6}/°C$

α_m = the thermal coefficient of linear expansion of the micrometer = $6.2 \times 10^{-6}/°C$
τ = the actual average ambient temperature = 23 °C
Δ_T = the deviation of the ambient measurement temperature from the reference temperature of 20 °C = +3 °C

Let the average value of five measurements be 20.005 mm.
The change in length due to the temperature difference is given by

$$\Delta x = x_n(\alpha_g - \alpha_m)\Delta T$$
$$= 20 \times (11.5 - 6.2) \times 10^{-6} \times 3$$
$$= 0.000\,318 \text{ mm}$$
$$= 0.318 \text{ μm.}$$

(The unit here is 'micrometer' and not 'micron.' See chapter 32, 'Analyzing the results,' for further discussion of this point.)
Hence, the corrected length at 20 °C is = 20.005 mm − 0.000 318 mm = 20.0046 mm.

(Incidentally, if rounded to the actual resolution of the micrometer, the result after correction is 20.005, which is same as that without correction!)

Such corrections, i.e. insignificant corrections, may not be applied in future if the measurement conditions remain the same. Nonetheless, this has to be an *informed decision.*

Note, however, that even if correction is neglected, its uncertainty must be taken into account. The uncertainty of this correction may be estimated by considering the following factors:

- error in temperature measurement,
- error due to ambient temperature variations,
- errors due to the actual temperatures of the gauge block and the micrometer, and
- errors due to the given values of the thermal coefficients of expansion of the gauge block and the micrometer

References

[1] JCGM 100:2008 *(GUM) Evaluation of Measurement Data—Guide to the Expression of Uncertainty in Measurement* 1st edn (BIPM, IEC, IFCC, ISO, IUPAC, IUPAP, OIML—International Organization for Standardization)
[2] JGCM 200:2012 *International Vocabulary of Metrology—Basic and General Concepts and Associated Terms* 3rd edn (Sèvres: BIPM)
[3] EUROLAB 2007 *Measurement Uncertainty Revisited: Alternative Approaches to Uncertainty Evaluation* (Paris: EUROLAB) Technical Report 1/2007

Chapter 27

The test uncertainty ratio: use only as a guiding phenomenon

Abstract: The concept of the test uncertainty ratio (TUR) has been quite popular among the metrology community for a couple of decades. Despite being declared redundant in today's metrology, it is still ingrained in the psyche of its practitioners.

> *Like other occult techniques of divination, the statistical method has a private jargon deliberately contrived to obscure its methods from non-practitioners.*
>
> —G O Ashley

The concept of the TUR has been prevalent in the metrology community for quite some time. Prior to that, the concept of the test accuracy ratio (TAR) dominated the field of metrology. The TUR still dominates when deciding whether a unit under test (UUT) can be accepted for testing or calibration vis-à-vis a standard's uncertainty.

A fact one should know

It is worth noting that neither the GUM [1] nor ISO/IEC 17025 [2] categorically mention this concept. JCGM200 (or VIM-International Vocabulary of Metrology) also does not mention these terms. However, ISO/IEC 17025:2017 says in clause 6.4.5: 'The equipment used for measurement shall be *capable* of achieving the measurement accuracy and/or measurement uncertainty required to provide a valid result.' Among other guides, only NATA mentions it in in appendix IX of [3], 'Assessment of Compliance with Specifications.'

TAR: The TAR originated in the 1960s in MIL-STD-45662A [4] as a crude way of assuring quality of measurement and reducing the 'false-accept risk' (FAR).

The primary objective was to have a simple criterion to ensure that the measurement standard had least influence on overall measurement process uncertainty. The TAR (and then the TUR) got a boost once it was adopted by standards such as MIL-STD-45662A for defense supplies and ANSI Z540 [5] for industrial supplies and became a de facto requirement in practically all fields of metrology. (The MIL standard did not originally use terms such as the TAR or TUR. In clause 5.2, under 'Adequacy of Measurement Standards,' it just says that the 'collective uncertainty of measurement standard shall not exceed 25% of the acceptable tolerance.') Both the TAR and the TUR were/are regarded as 'figures of merit' for the selection of standards for a particular measurement.

The TAR is defined as the ratio of the UUT tolerance to the accuracy of the measurement standard. (The accuracy may be specified as the standard's tolerance or uncertainty at a certain confidence interval, the latter being preferred.) Hence

TAR = UUT tolerance/measurement standard tolerance.

The TAR was in vogue in industry for a long time but subsequently lost its appeal when it was realized that, in addition to the measurement standard's tolerance, other factors contribute to overall uncertainty. This overall uncertainty is nothing but 'measurement process uncertainty,' which is termed 'expanded uncertainty' in the GUM. Thus, the TAR was replaced by the TUR. (ANSI/NCSL Z540.3 does not consider or use the TAR [5].)

The TUR: the TUR is defined as the ratio of the UUT tolerance (specifications) to the uncertainty of the measurement process. ANSI/NCSL Z540.3 defines it as a 'ratio of the span of the UUT tolerance limits to twice the 95% expanded uncertainty of the measurement process.' Hence,

$$\text{TUR} = \frac{L_1 + L_2}{2U_{95}} \tag{27.1}$$

where L_1 and L_2 are the lower and upper tolerances (specifications) of the UUT.

In ANSI/NCSL Z540, the coverage factor is fixed at a 95% confidence level, $k_{95} = 2$, and the TUR is valid only when $L_1 = L_2 = L$. Therefore, effectively,

$$\text{TUR} = \frac{L}{U_{95}}. \tag{27.2}$$

Clause 5.3 of ANSI/NCSL Z540.3 states: 'Where calibrations provide for verification that measurement quantities are within specified tolerances, the probability that incorrect acceptance decisions (false accept) will result from calibration tests shall not exceed 2% and shall be documented. Where it is not *practicable* to estimate this probability, the test uncertainty ratio shall be equal to or greater than 4:1.'

Historically, a TUR of 10:1 was a rule of thumb that was gradually scaled down to 4:1. Basically, the purpose of the TUR is to control the probability of false accept (PFA) or consumer's risk by way of keeping a tab on measurement process uncertainty. See table 27.1 that shows risks at TURs of 4:1 and less.

ANSI/NCSL Z540.3 marked a paradigm shift in judging the calibration process capability. The decision process changed from the TAR/TUR to the PFA in order to

Table 27.1. The false-accept risks (FARs) & false-reject risks (FRRs) at various TURs at specification limit $= 2\sigma$ (at a 95% CL).

TUR	Approx. false-accept risk (FAR) %	Approx. false-reject risk (FRR) %
4:1	0.8	1.3
3:1	1.0	2.1
2:1	1.1	4.2
1.5:1	1.3	7.0
1:1	1.8	11.0

protect the customer. ANSI/NCSL Z540.3 specified a limit on the PFA of no more than 2%. A TUR of $\geq$ 4:1 was supposed to keep the PFA at an acceptable level, but ANSI/NCSL Z540.3 explicitly mentioned that the TUR should be used if it was *impracticable* to calculate the PFA. In other words, if the PFA was within 2%, the TUR was not an issue!

A fact one should know

Note that the TUR and the TAR are different by definition and should not be swapped as if they are synonymous. Their values will be close only if the measurement process uncertainty is dominated by the standard's uncertainty, i.e. other factors make an insignificant contribution to measurement process uncertainty.

The measurement process uncertainty (MPU) can be shown to be $\left[\sqrt{(1 + TUR^2)}/TUR\right]$ times the UUT uncertainty. Based on this relationship, table 27.2 shows how the TUR affects the MPU.

Table 27.2. The effect of the TUR on measurement process uncertainty.

TUR	$\sqrt{(1 + TUR^2)}/TUR$	% increase in MPU
10:1	1.005	0.5
5:1	1.020	2.0
4:1	1.031	3.1
3:1	1.054	5.4
2:1	1.118	11.8
1:1	1.414	41.4

Thus, a TUR $\geq$ 4:1 ensures that the standard does not contribute to overall uncertainty by more than about 3%. Remember, however, that this does not endorse FAR mitigation to any specific level.

A critique of the TUR: a TUR of 4:1 essentially became a doctrine for the selection of a measurement standard. In absence of state-of-the-art tools in former days, it was seen as a simple, cost-effective, and supposedly reliable figure of merit that safeguarded the interests of all stakeholders. However, with the advent of software tools and advanced computing power that complemented statistical analysis, it was realized that the TUR was not a robust criterion in regard to the PFA. Some of the objections to the TUR are:

1. It is a crude risk-control criterion and does not control the risk to any specific level.
2. If the UUT is already in tolerance, the FAR is zero irrespective of the TUR. To estimate the PFA, the *a priori* in tolerance probability of the UUT is required to be known, and the TUR is not an affirmative marker of that probability.
3. It is limited to two-way symmetric tolerances only ($L_1 = L_2$).
4. The definition (given in ANSI/NCSL Z540.3) accounts for expanded uncertainty at a 95% confidence level only.

So, when should one resort to the 4:1 TUR criterion? It should be used only when:

- a simple criterion is required irrespective of FARs
- the *a priori* in tolerance probability is unknown
- the calculation of the PFA or FAR is impracticable, but one wants to comply with ANSI/NCSL Z540.3 requirements.

A TUR of $\leq$ 3:1 is also accepted these days, as the cost of high-precision UUTs and industrial equipment is becoming affordable, but the overall cost of maintaining standards that meet a TUR of $\geq$ 4:1 (including traceability and keeping standby standards) is prohibitively high.

A fact one should know

JCGM 106:2012 [6] uses a term 'measurement capability index,' C_m, defined as

$$C_m = \frac{T_U - T_L}{4u_m}$$

where T_U and T_L are the upper and lower tolerance limits of the UUT and u_m is the combined uncertainty of measurement. The factor of four in $4u_m$ is arbitrary, and when a coverage factor of two at a 95% confidence level is used, $4u_m$ reduces to $2U$, where U is the expanded uncertainty. With symmetric tolerance, the above equation exactly resembles the equation for the TUR as specified in ANSI/NCSL Z540.3 (at a coverage factor of two and a 95% level of confidence)

JCGM 106:2012 acknowledges that parameters such as the TUR, gauging ratio, TAR, etc. have a close connection with C_m but advises the reader to take care when using these parameters. On the other hand, it claims that while the above parameters are 'ambiguous,' C_m is clearer!

C_m and the TUR resemble the process capability index C_p used in statistical process control (SPC) and Six Sigma.

References

[1] JCGM 100:2008 *(GUM) Evaluation of Measurement Data—Guide to the Expression of Uncertainty in Measurement* 1st edn (BIPM, IEC, IFCC, ISO, IUPAC, IUPAP, OIML—International Organization for Standardization)

[2] ISO/IEC 17025:2017 *General Requirements for the Competence of Testing and Calibration Laboratories* (Geneva: ISO)

[3] Cook R R 2002 *Assessment of Uncertainties of Measurement for Calibration and Testing Laboratories* (Rhodes, NSW: NATA)

[4] MIL-STD-45662A 1988 *Military Standard: Calibration System Requirements* (Washington, DC: US DoD) NOTE: This Standard is Replaced by ISO 10012 [Reviewed and Confirmed in 2022]

[5] ANSI/NCSL Z540.3-2006 *Requirements for the Calibration of Measuring and Test Equipment, National Conference of Standards Laboratories* (Washington, DC: ANSI)

[6] JCGM 106:2012 *Evaluation of Measurement Data—The Role of Measurement Uncertainty in Conformity Assessment* (Sèvres: BIPM)

IOP Publishing

A Practical Handbook on Measurement Uncertainty
FAQs and fundamentals for metrologists
Swanand Rishi

Chapter 28

Guarding conformity decisions

Abstract: The use of a guard band (GB) reduces the probability of making an incorrect conformance decision. It is basically a safety factor built into the measurement decision process (pass/fail) that reduces the acceptance limit below the specification/tolerance limit. Measurement uncertainty is widely accepted measure of the GB—directly, or as a multiple of it.

> *Statistics suggest that when customers complain, business owners and managers ought to get excited about it. The complaining customer represents a huge opportunity for more business.*
>
> —Zig Ziglar

In many situations, we are expected to check the conformity of a measurement against specifications. Until the time uncertainty was no longer a mandatory requirement, the decisions about conformity were simple and straightforward, i.e. if the measured value was within the specification limits, conformity was met and the product was accepted; otherwise not.

The picture changed when standards related to measurement required that the uncertainty of measurement must be taken into account for conformity decisions. Clause 7.8.6.1 of ISO/IEC 17025:2017 [1] states: 'When a statement of conformity to a specification or standard is provided, the laboratory shall document the decision rule employed....' Here, the decision rule, as defined in clause 3.7, is a 'rule that describes how measurement uncertainty is accounted for when stating conformity with a specified requirement.' The reason is that the uncertainty is the doubt about the measurement and hence must be accounted for when stating conformity, i.e. pass/fail decisions.

doi:10.1088/978-0-7503-6462-1ch28

A fact one should know

As per clause 3.7 of ISO/IEC 17025:2017, a decision rule is a rule that describes how measurement uncertainty is accounted for when stating conformity with a specified requirement.

Clause 7.1.3 of ISO/IEC 17025:2017 states that 'Unless inherent in the requested specification or standard, the decision rule selected shall be communicated to, and agreed with, the customer. This means that if the decision rule is implied or if it is part of a specification or standard, there is no need to specify the decision rule separately.

Elaborate guidance on statements of conformity is provided in ISO/IEC Guide 98–4 [2]. ILAC-G8:09:2019 also gives examples of statements of conformity in appendix B [3].

As uncertainty is given with upper and lower limits, it creates peculiar overlapping situations in decision-making, as explained later in figure 28.1.

Facts one should know

The terms conformity, compliance, and conformance are often used interchangeably. However, these terms are not synonymous. ISO/IEC 17025:2005 used the term 'compliance' (not conformity), whereas its 2017 edition uses term 'conformity' (not compliance). Compliance basically means meeting the requirements of a regulation created by an authority (e.g. by government) or that of a statutory body (established by authority). *The term conformance is no longer used to mean conformity.*

ISO/IEC 17025:2017 recognizes this fact but provides no specific *guidance* about *how* to take the measurement uncertainty into account when assigning pass/fail status.

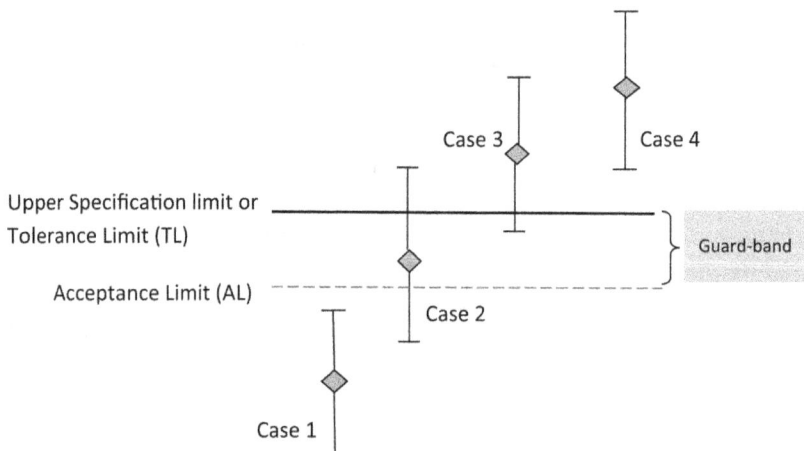

Figure 28.1. A GB (shown $\cong$ at uncertainty).

However, it is important that the decisions are correct and not biased towards either the supplier or the customer. In order to have some assurance, the practice of providing a so-called GB was adopted. GBs are also employed to reduce the false-accept risk (the Consumer's risk) as well as the false-reject risk (the Producer's risk).

These aspects of the GB and risks are elaborately discussed in ISO/IEC Guide 98–4 and JCGM 106.

As per ILAC-G8:09/2019, there are two types of risks:

Specific risk: this is the probability that an accepted item is non-conforming, or that a rejected item does conform. This risk is based on measurements of a single item.

Global risk: this is the average probability that an accepted item is non-conforming, or that a rejected item does conform. It does not directly address the probability of false acceptance of any single item, discrete measurement result, or individual workpiece. A global risk of $< 2\%$ is common, as per ANSI Z540.3.

28.1 What is a guard band?

ASME B89.7.3.1-2001 [4] (Revised 2019) defines the GB as 'the magnitude of the offset from the specification limit to the acceptance or rejection zone boundary' while in JCGM 106, the GB is 'the interval between a tolerance limit and a corresponding acceptance limit.'

So, a GB is nothing but a range by which the specification limit (SL) is reduced to what is called an 'acceptance limit' (AL), so that there is more confidence in pass/fail decisions. Conformity decisions are taken based on tolerance limits rather than specification limits. Thus, the GB provides a safety factor. It is particularly employed when the TUR is lower than 4:1 and is a safety margin for deciding an acceptance (pass) limit. Thus, the GB (w) is the interval between a tolerance limit and a corresponding acceptance limit, where length $w = |TL - AL|$.

An SL (or tolerance limit (TL)) is the specified upper or lower bound of the permissible values of *a property*, whereas an AL is the specified upper or lower bound of the permissible *measured* quantity values. Thus, the guard band, denoted by w, equals the TL minus the AL:

$$w = (TL - AL) \tag{28.1}$$

It can also be expressed as a multiple of the TL, i.e.

$$GB = w = K \times TL, \tag{28.2}$$

where $K \leqslant 1$; its value depends upon the strategy adopted by the decision maker.

SLs are specified by the manufacturer either as a rectangular, usually two-sided symmetrical specification, or at a 95% level of confidence. It is stated in the GUM that when a specification is quoted for a given coverage probability, then a normal (Gaussian) distribution can be assumed.

As ISO/IEC 17025:2017 does not specify a maximum level of false-accept risk, it allows us to follow different guard banding strategies (although it does not explicitly mention the words 'guard band'). The simplest among them is to consider the GB equal to the (expanded) uncertainty. Clause 5.2 of ILAC-G8:09:2019 states that it is common to use a GB of $w = U$.

Figure 28.1 illustrates the concept given in ILAC-G8:09/2019 (clause 4.2.3 for non-binary statement with GB). The GB in the figure is approximately set at the expanded uncertainty (at a 95% level of confidence) for the sake of illustration. (In ILAC-G8:09, this figure is shown without GB and pass/fail legends.) It is shown for an upper specification limit but also applies to lower specification limits.

Table 28.1. Applying uncertainty in conformity decisions.

Parameter: DC voltage

Ser. no.	Standard value (+V)	Spec. limit (SL) (+V)	Measured value (+V)	Acceptance or tolerance limit (TL) (+V)	Expanded uncertainty (+/−V)	Status*
1	10.0000	10.0015	10.0009	10.0013	0.0005	Pass
2	10.0000	10.0015	10.0011	10.0013	0.0005	Pass^
3	10.0000	10.0015	10.0014	10.0013	0.0005	Fail^
4	10.0000	10.0015	10.0019	10.0013	0.0005	Fail^
5	10.0000	10.0015	10.0021	10.0013	0.0005	Fail

The ^ mark shows cases in which an explanation shall be given.

The diamonds in the figure are different measurement results with uncertainty bands above and below them.

Table 28.1 shows how to account for uncertainty in conformity decisions for non-binary statements with GBs. Also refer to figure 28.1 for decisions regarding the pass, fail, pass^, and fail^ statuses shown in the last (status*) column of table 28.1.

A table without stars is like champagne without bubbles!

—David Giles

A GB of 0.0002 V is considered in this table, hence TL = 10.0015 V−0.0002 V = 10.0013 V. The decisions denoted by pass^ and fail^ statuses are worth noting, as they represent gray areas and should be clearly and explicitly reported as explained under (c) above.

Fact one should know

Note that the use of GB does not altogether avoid a false-accept or false-reject risk, but provides a safety margin and increases confidence in pass/fail decisions. Various methods (each using a different formula and basis) are adopted to decide upon a GB. Many of these use the TUR in the formula used to calculate the TL. Depending upon the guard banding technique chosen, both the false-accept and false-reject risks vary considerably.

There are a few other guard banding methods, as listed below. (The list is not comprehensive and is mentioned for reference only.)

1. UKAS M3003:2007; appendix M, Clause M2 [5]
2. UKAS M3003:2007; appendix M, Clause M3
3. The RSS method
4. The NCSL Recommended Practice 10 (RP-10) method [6]
5. The MIL-STD-45662A and ANSI Z540.1 method [7]
6. The managed risk method by Michael Dobbert of Agilent Technologies (for meeting the ANSI/NCSL Z540.3:2006 [8] requirement for a false accept probability of $\leqslant 2\%$) [9].

Although the concept of the GB had roots in safeguarding measurements done with TURs of less than 4:1, it is now gaining momentum as a methodology for controlling false-accept risks. All the methods mentioned above are pertinent but should be used in the right context. The method to be followed should be an informed decision that is taken with the customer in the loop. The significance, criticality, and reliability of the application (e.g. space, military, or industrial) as well as the financial implications should be factored in when adopting any of the methods.

The ASME B89.7.3 working group [4] states that the selection of a decision rule is a *business* decision and that the flexibility of having a continuum of rules ranging from stringent to relaxed acceptance or rejection is needed in order to satisfy a broad range of industries.

References

[1] ISO/IEC 17025:2017 *General Requirements for the Competence of Testing and Calibration Laboratories* (Geneva: ISO)

[2] ISO/IEC Guide 98-4:2012 *Uncertainty of measurement Part 4: Role of measurement uncertainty in conformity assessment* (Geneva: ISO)

[3] ILACG8:09/2019 *Guidelines on Decision Rules and Statements of Conformity* (Adelaide: ILAC)

[4] ASME B89.7.3.1-2001 *Guidelines for Decision Rules: Considering Measurement Uncertainty in Determining Conformance with Specifications* (New York, NY: ASME) [Reaffirmed in 2019]

[5] UKAS M3003 2007 *The Expression of Uncertainty and Confidence in Measurement, United Kingdom Accreditation Service* (Staines-upon-Thames: UKAS)

[6] NCSL International 2021 *RP-10: Establishment and operation of an electrical utility metrology laboratory* (Lafayette, CO: NCSL International)

[7] MIL-STD-45662A 1988 *Calibration System Requirements* (Washington, DC: US DoD) NOTE: This standard is Replaced by ISO 10012:2003 Measurement Management Systems Requirements for Measurement Processes and Measuring Equipment [Reviewed and Confirmed in 2022]

[8] ANSI/NCSL Z540.1-1994 (R2002) *Calibration Laboratories and Measuring and Test Equipment—General Requirements* (Washington, DC: ANSI) [Withdrawn as an active standard from July 2007 and superseded by ANS/ISO/IEC 17025:2005 for part 1 and ANSI/NCSL Z540.3–2006 for part 2. ISO/IEC 17025:2005 is now ISO/IEC 17025:2017]

[9] Dobbert M 2008 A Guard-Band Strategy for Managing False-Accept Risk *NCSLI Measure* **3** 44–8

Part F

Delving a little deeper

Chapter 29

Treating dominant non-Gaussian components

Abstract: Although the underlying assumption in the GUM [1] is a Gaussian or Normal distribution, the evaluation of uncertainty does involve other types of distributions. A problem arises when metrologists have to face a dominant contribution to uncertainty that has a non-Gaussian distribution. This needs to be treated carefully.

> *Statistics are no substitute for judgment.*
>
> —Henry Clay

The underlying assumption in the GUM that is used to model the mathematical treatment is a Gaussian or Normal distribution. As indicated many times in the GUM, this distribution can be assumed in most practical cases. It is also a cardinal or principal requirement of the central limit theorem (CLT) and the law of propagation of uncertainty (LPU).

The significance of the Normal distribution: this distribution was invented by Abraham De Moivre in 1773; however, it is named after Gauss because he first tried to apply it in scientific measurements of astronomical objects and is thus credited for popularizing it. It is the most common probability distribution due to its unique properties; a few of its important properties were already discussed in chapter 11, 'The Normal distribution, the t-distribution, and the standard Normal distribution.' However, they are listed again here for convenience.

1. The Normal distribution is a unimodal distribution that has a bell-shaped curve. It has the same mean, mode, and median (which are measures of central tendency).
2. It is a symmetric continuous distribution. Due to this and (1) above, the areas under the curve on either side of the ordinate at $X = \mu$ (where μ is the mean) are equal.
3. The X-axis is asymptotic to this distribution, which extends from $-\infty$ to $+\infty$.
4. Its PDF can be defined by its mean μ and its standard deviation σ. (No additional parameters such as shape, location, etc are required to define it.)

doi:10.1088/978-0-7503-6462-1ch29

5. Most of the data in economic and business statistics adhere to this distribution. It is also observed in many industrial, trade, and commercial applications and hence is very realistic. This characteristic makes it the most useful distribution. (This can be compared with the sinusoidal waveform, which has many more applications in electrical engineering, mechanics, and motion studies than other types of waveforms.)

6. Many other distributions (continuous as well as discrete) approach the Normal distribution under limiting conditions, the significant condition being a large sample size (i.e. the sample size 'n' tends to ∞).

7. It is also produced by convolving other distributions, albeit under some conditions. This characteristic is the basis of the CLT, which is the backbone of the GUM.

As the underlying assumption in the GUM is the presence of a Gaussian or Normal distribution, a problem arises when metrologists have to face a dominant contribution to uncertainty that has non-Gaussian distribution.

There are no clear-cut rules to determine the dominance of a non-Gaussian distribution, but certain guidelines are given in some guides and other publications. The typical guidelines given in appendix C of UKAS M3003:2007 [2] are as follows:

- If the expanded uncertainty calculated by the usual formula (RSS) at a coverage factor of $k = 2$ is greater than the arithmetic sum of the limit values of all components, this indicates the likelihood of a dominant non-Gaussian distribution, generally a rectangular distribution or U (arcsine) distribution.

- In the case of a dominant component with a Normal distribution, the limit value is taken to be three times the standard deviation.

In such cases, UKAS M3003:2007 indicates that the expanded uncertainty should be expressed as a two-part expression, as follows:

$$U = U' + a_d,$$

where U' is the expanded uncertainty calculated as usual but *excluding* the dominant factor a_d. UKAS M3003:2007 anticipates that this dominant factor a_d is not likely to be dominant when it is used in the subsequent evaluation of uncertainty and can be assumed to have a Normal distribution. (However, this assumption is naïve and needs to be verified practically in each case. If it is not valid, no alternative method is suggested in UKAS M3003:2007, and one will have to forgo the two-part expression and instead revisit the calculation to get a single value. It should be noted that the 2023 edition of UKAS M3003 does not discuss the two-part expression discussed in the 2007 edition.)

Where a single value of uncertainty is required (as recommended in the GUM), and we have to convolve different distributions, appendix C of UKAS M3003 gives suitable coverage factors at a 95.45% coverage probability. (The GUM does not give such a table of values for coverage factors for different convolved distributions.) Table 29.1 (reproduced from UKAS M3003:2007/2023) shows the coverage factor k at a coverage probability of 95.45% for the most common convolution scenario, i.e. the convolution of a rectangular distribution and a Normal distribution.

Table 29.1. The coverage factor k at a coverage probability of 95.45% for the convolution of a rectangular distribution and a Normal distribution (Ref: UKAS M3003:2023, appendix C, clause No. C7). Reproduced from [2] with permission from the United Kingdom Accreditation Service (UKAS).

$U_i(y)_{\text{Normal}}/U_i(y)_{\text{rect}}$	$k_{95.45\%}$	$u_i(y)_{\text{Normal}}/U_i(y)_{\text{rect}}$	$k_{95.45\%}$	$u_i(y)_{\text{Normal}}/U_i(y)_{\text{rect}}$	$k_{95.45\%}$
0.00	1.65	0.50	1.84	0.95	1.95
0.10	1.66	0.55	1.85	1.00	1.95
0.15	1.68	0.60	1,87	1.10	1.96
0.20	1.70	0.65	1.89	1.20	1.97
0.25	1.72	0.70	1.90	1.40	1.98
0.30	1.75	0.75	1.91	1.80	1.99
0.35	1.77	0.80	1.92	2.00	1.99
0.40	1.79	0.85	1.93	2.50	2.00
0.45	1.82	0.90	1.94	∞	2.00

A fact one should know

UKAS M3003:2007 (and also the 2023 edition) gives a simple test (as a rule of thumb) that can be used to determine whether a rectangular distribution is a dominant component. If standard uncertainty of the dominant component is more than 1.4 times the combined standard uncertainty for the *remaining* components, it is a dominant component!

For example, if the standard uncertainty x_1 of a component that has rectangular distribution is 5% and the combined standard uncertainty x_2 of the remaining components is 3.3%, then since 5% > 3.3% * 1.4, x_1 can be considered to be a dominant component.

If it is not, then $u_i(y)_{\text{Normal}}/U_i(y)_{\text{rect}}$ is $\geqslant 0.71$ and the coverage factor k will be within 5% of the usual value of 2.00. (Note from the above table that $k = 1.90$ at a $u_i(y)_{\text{Normal}}/U_i(y)_{\text{rect}}$ value of 0.70. Thus, the deviation of the coverage factor (and hence the expanded uncertainty) from $k = 2$ is within 5% if $u_i(y)_{\text{Normal}}/U_i(y)_{\text{rect}}$ is $\geqslant 0.71$.)

Some more considerations:

1. In some special cases, if some components are known to contribute to uncertainty but cannot be quantified or assessed with rigor, this fact may be stated with the result.

For example, in vernier caliper calibration, the effects of flatness error are known to contribute to uncertainty. However, if the data for the flatness error is not available, then this could be stated in the report.

2. Among various distributions, a uniform distribution is most likely to be dominant. In such cases, the estimated expanded uncertainty is conservative but most likely to be unrealistically large.

> For example, for a semi-width of a and coverage factor $k = 2$, the expanded uncertainty is $2a/\sqrt{3} = 1.15a$, meaning it is 15% larger than the known dominant semi-width a. Although this is theoretically sound, it poses some difficulty in making pass/fail decisions with respect to specification limits.

Clarifications of such facts need to be given when reporting the results.
3. As a rule of thumb, if a contribution is more than twice the next largest contribution, it may be treated as dominant. Some metrologists consider that if there are three contributors and if largest of them is more than three times the next largest, it can be treated as a dominant contribution.

References

[1] JCGM 100:2008 *(GUM) Evaluation of Measurement Data—Guide to the Expression of Uncertainty in Measurement* 1st edn (BIPM, IEC, IFCC, ISO, IUPAC, IUPAP, OIML—International Organization for Standardization)
[2] UKAS M3003 2007 (revised/reviewed and confirmed in 2023) *The Expression of Uncertainty and Confidence in Measurement* (Staines-on-Thames: UKAS)

IOP Publishing

A Practical Handbook on Measurement Uncertainty
FAQs and fundamentals for metrologists
Swanand Rishi

Chapter 30

Sample analysis—how normal is the Normal?

Abstract: For the Type A evaluation method, the sample data is assumed to be normally distributed even if the sample size is small. This assumption may prove to be wrong. There are some tests that can be used to check the assumption of normality of the collected measurement data.

> *There are two ways of lying. One, not telling the truth and the other, making up statistics.*
>
> —Josefina Vazquez Mota

As discussed elsewhere, a Normal distribution is the underlying distribution of the GUM [1] uncertainty framework (GUF). Its characteristics are also discussed at length in chapter 29, 'Treating dominant non-Gaussian components.' The distribution of the combined uncertainty is also assumed to be a χ^2 distribution (that approaches a Normal distribution for large n) if the conditions of the central limit theorem (CLT) and the law of propagation of uncertainty (LPU) are fulfilled.

In most statistical analyses of sampled data, it is assumed that the distribution of data is normal or approximately so, provided the sample size is large. The conclusions of analysis are very much dependent on this assumption, and hence it is necessary to assess the normality of data. In practice, we deal with a limited sample size that ranges from three to ten measurements. As such, we are not sure whether the sample data really represents the population, and we are confronted with a question: 'How normal is the normal?'

Normality testing: in order to check the normality of the data, various tests are available. Normality can be assessed by qualitative or quantitative tests.

> *If I can't picture it, I can't understand it*
>
> —Albert Einstein

doi:10.1088/978-0-7503-6462-1ch30

1. **Qualitative tests:** these are graphical methods that mainly include histograms and normal probability plots. Qualitative tests are easy and provide good and concise visual depiction, but their interpretation is subjective.

   - **Histograms:**

        The histograms in figures 30.1 and 30.2 depict normal and non-normal data respectively, while figures 30.3 and 30.4 illustrate normal probability plots for normal and non-normal data, respectively. (Plots can be drawn using software tools or a spreadsheet. The trend line has been shown for illustration purposes only.)

   - **Normal probability plots:**

A fact one should know

To form a more objective view of a normal probability plot, one can find the value of the test statistic of the observed data and compare it with the significance level (also called the α-risk). If value of the test statistic is less than the α-risk, the data is not normal. For a common confidence level of 95%, the α-risk is $(1-0.95) = 0.05$. Hence, if the value of the test statistic is less than 0.05, the data is not normal.

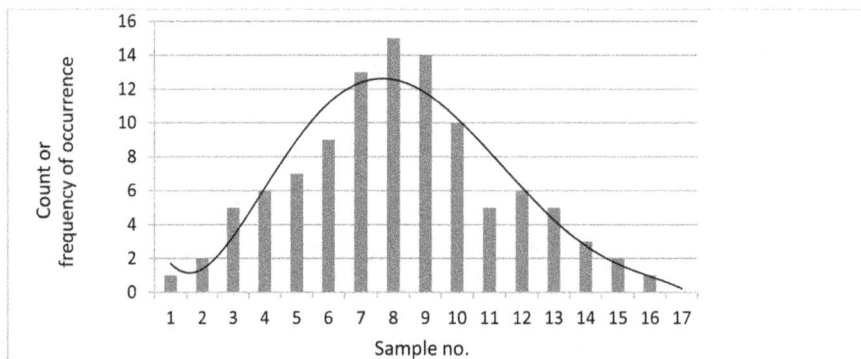

Figure 30.1. A histogram indicating fairly good normality of data.

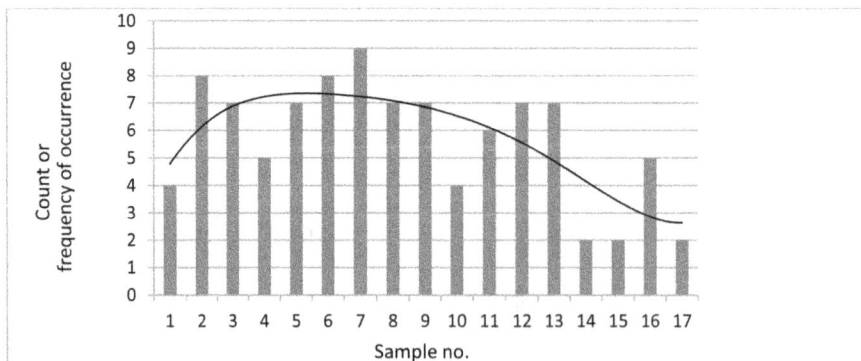

Figure 30.2. A histogram indicating non-normality of data.

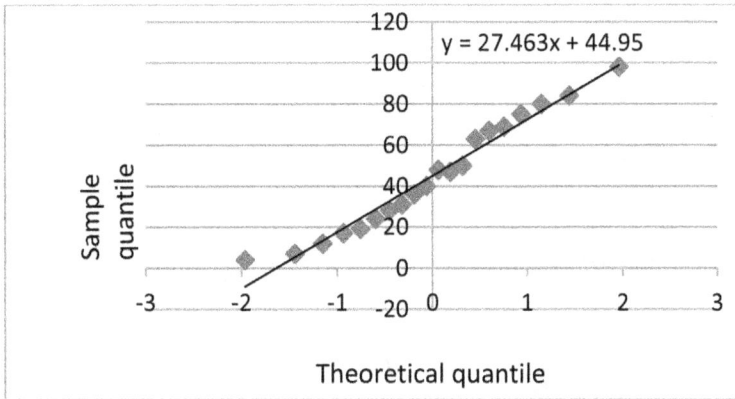

Figure 30.3. A normal probability plot indicating 'normality' of data.

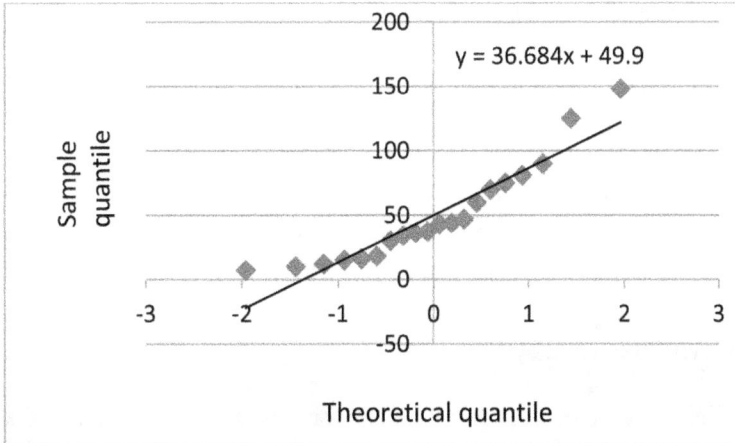

Figure 30.4. A normal probability plot indicating 'non-normality' of data.

Qualitative tools are generally not suitable if the sample size is less than ten. These tests provide correct information if:
- the model is properly defined,
- the standard deviation is constant throughout the data,
- the process is fairly stable, and
- random errors are independent from one run to another.

A fact one should know

Both histograms and normal probability plots can be constructed in Excel spreadsheets using column charts and scatter charts, respectively. The use of the histogram is quite straightforward. For a scatter chart, arrange the data in ascending order. Find the cumulative probability using [(sample number – 0.5)/sample size] (a common formula) for each reading. Then, find the normal score for each cumulative probability value using the NORMSINV function. Select all the cells corresponding to 'normal score' and 'data' to obtain a scatter plot.

Numerical quantities focus on expected values, graphical summaries on unexpected values.

—Tukey

2. **Quantitative tests:** these include 'skewness' (a measure of data symmetry) and 'kurtosis' (a measure of data 'peakedness') tests, the 'chi-square test,' and other tests named after statisticians. The quantitative tests are more involved but provide objective information.

- **The coefficient of skewness:** designated by c_3, it is given by

$$c_3 = \left[(1/n)\sum_{i=1}^{n}(x_i-\bar{x})^3 \right]/s^3. \tag{30.1}$$

- **The coefficient of kurtosis:** designated by c_4, it is given by

$$c_4 = \left[(1/n)\sum_{i=1}^{n}(x_i-\bar{x})^4 \right]/s^4, \tag{30.2}$$

with the usual symbols.

The coefficients of skewness and kurtosis for a Normal distribution are zero and three, respectively. A significant deviation from these figures (dependent on application and criticality) indicates non-normal data.

A positively or right-skewed distribution has large values at its extremes, while a negatively or left-skewed distribution has small values at its extremes. Distributions whose kurtosis is much higher than three have heavier tails and most likely contain outliers.

The chi-squared goodness-of-fit test: The χ^2 test is suitable for sample sizes > 50 and is based on the relative difference between observed frequencies and the theoretical frequencies predicted by the probability density function of the Normal distribution. A popular statistic for this test is Pearson's chi-square statistic, given by

$$\sum_{i=1}^{n}\frac{(O_i - E_i)^2}{O_i}, \tag{30.3}$$

where O_i is the ith observed value (the inputs are assumed to be independent quantities) and E_i is the corresponding expected value (E_i is generally a single reference value).

The p-value obtained for the sample data from equation (30.3) is compared with the significance level α (which for a 95% confidence level is 0.05), and if it is more than the latter, normality can be assumed.

- **The Shapiro–Wilk test:** this test is generally used for sample sizes between 20 and 50, and it computes a W-statistic given by

$$W = \left(\sum_{i=1}^{n} (a_i y_i) \right)^2 / SS,$$
(30.4)

    where n is the sample size, a_i are the Shapiro–Wilk coefficients (available in handbooks and given in appendix A), y_i are the input quantities, and $SS =$ the sum of the squared differences, $\sum_{i=1}^{n} (x_i - \bar{x})^2$.

For calculation, sample data is arranged in ascending order and named as x_i and the above equation becomes,

$$W = \left(\sum_{i=1}^{m} a_{n+1-i}(x_{n+1-i} - x_i) \right)^2 / SS,$$
(30.5)

where $m = n/2$ if n is even and $m = (n - 1)/2$ if n is odd. (If n is odd, the median data value is not used.)

For sample size n, a critical value W_c that is closest to W is found (by interpolation if necessary) from the Shapiro–Wilk table. For this W value, a corresponding p-value is obtained. If this p-value is greater than the significance level α, the assumption of normality can be accepted. Usually, we evaluate uncertainty at a 95% level of confidence, for which the significance level α is 0.05. Hence, if the p-value is > 0.05, normality can be assumed. Although this test is suitable for sample sizes between 20 and 50, coefficients a_i are available from $n = 2$ (see appendix A).

A fact one should know

The Shapiro–Wilk normality test is also sensitive to small sample sizes ($n < 20$), asymmetry of distribution, and long or short tails. However, for large n, the calculations are cumbersome due to almost half a dozen coefficients and interpolation. Apart from this drawback, the test has proved quite versatile and is well accepted in the metrology community.

Let us see an example. We will use the data for the pressure measurement results in the Excel sheet below. Ten pressure readings are taken, as shown in column B.

Note: table A-1 and table A-2 mentioned in the example and in the Excel sheet refer to the respective tables in appendix A.

The Shapiro–Wilk test for normality: pressure in kg . m^{-2}.

Ser. no.	Quantity reading	Readings in ascending order	Weights a_i	Value of a_i	Difference	value	$b_i = a_i \times$ diff
1	25.8	22.9	a_1	0.5739	$x_{10}-x_1$	4.4	2.525 16
2	27.2	23.6	a_2	0.3291	x_9-x_2	3.6	1.184 76
3	24.9	24.9	a_3	0.2141	x_8-x_3	1.2	0.256 92
4	22.9	24.9	a_4	0.1224	x_7-x_4	0.9	0.110 16
5	25.3	25.3	a_5	0.0399	x_6-x_5	0.2	0.007 98
6	26.1	25.5					
7	27.3	25.8					
8	24.9	26.1					
9	25.5	27.2					
10	23.6	27.3					

In this example,
$n = 10$
$SS = 17.485$
$\sum b = 4.084\ 98$
$W = (\sum b)^2/SS =$ For n = 10, from table 2 of p values, this value lies between
0.954 364 $p = 0.5$ and 0.9
W for $p = 0.5$
0.938
W for $p = 0.9$ Slope =   11.7647
0.972
p-value 0.692 522
(for $W = 0.995\ 4364$ by
  interpolation)
     So, as $p = 0.692\ 522 > 0.05$, normality of data can be assumed.

The steps to be followed are given below with the values calculated in the Excel sheet.
- Arrange the data in ascending order.
- Calculate SS using

$$SS = \sum_{i=1}^{n}(x_i - \bar{x})^2.$$

Calculate b using

$$b = \sum_{i=1}^{m} a_i(x_{n+1-i} - x_i)^2.$$

- Refer to the a_i weights in table 1 (based on the value of n) in the Shapiro–Wilk tables. Calculate the test statistic $W = b^2/SS$.
- Find the value in table 2 of the Shapiro–Wilk tables (for a given value of n) that is closest to W. (Interpolate if the result is close to 0.05, which is a p-value

at a 95% confidence level, which is generally assumed.) This is the p-value for the data under test.

- If the p-value > 0.05, the data can be assumed to be normally distributed.
- In the given example, the normality of the data can be assumed because the p- value of 0.692 522 is > 0.05.

Reference

[1] JCGM 100:2008 *(GUM) Evaluation of Measurement Data—Guide to the Expression of Uncertainty in Measurement* 1st edn (BIPM, IEC, IFCC, ISO, IUPAC, IUPAP, OIML— International Organization for Standardization)

IOP Publishing

A Practical Handbook on Measurement Uncertainty
FAQs and fundamentals for metrologists
Swanand Rishi

Chapter 31

Sample analysis—detecting the outliers

Abstract: During practical observations and the collection of measurement data, we sometimes encounter some data points in a set that look weird. This raises the question of whether such a data point can be treated as an authentic reading or needs to be discarded. This chapter discusses this issue.

> *Strange events permit themselves the luxury of occurring.*
>
> —Charlie Chan

An outlier is an observed value that seems to be 'incompatible' with the other values observed within a sampled data set. Generally, in a controlled process, the outliers are few and far between, because they have a very low probability of belonging to the distribution. It is quite possible that the outlier is located near or beyond the distant part of the tail of the distribution. As we use standard deviation as a measure of variability, even one or two outliers can drastically affect its value or introduce a bias into the mean value. It is very difficult to identify an outlier from a small sample size as, in that case, a single outlier drastically changes the values of the mean and the standard deviation. Thus, it is essential to be able to identify them correctly. However, defining and thus identifying outliers is usually subjective and beyond consensus, hence their exclusion should be based on an adequate knowledge of the measurement process and should be an informed decision.

According to ASME B89.7.3.1-2001 [1], from a measurement point of view, an outlier must satisfy two conditions simultaneously: '(a) the anomalous reading cannot be repeated; and (b) the anomalous reading does not represent the system under test.'

ANSI ASTM E 178 [2] defines an outlier (also called an outlying observation) as 'an observation that appears to deviate markedly in value from other members of the sample in which it appears.'

A number of tests are available for identifying outliers from sampled data; these include Chauvenet's criterion and tests by Grubbs, Dixon, and Rosner. Chauvenet's criterion is widely accepted in metrology and hence elaborated here. It can identify one or more outliers from the sampled data.

Chauvenet's criterion: this defines an acceptable scatter of data around a mean value x for a given sample n and a standard deviation s_n. It specifies that all points should be retained that fall within a band around the mean value corresponding to a probability P_n of $(1 - (1/2n))$. This means that points can be considered for rejection only if the probability of obtaining their deviation from the mean is less than $1/(2n)$. The Normal distribution (with a typical 95% probability) is used to determine the number of sample standard deviations.

A doubtful reading x_d and its deviation d_d that may lie beyond a value given by $\tau_{max} s_n$ are rejected. Here, τ_{max}, which is the maximum interval around the mean, is given by

$$\tau_{max} = |x_d - \bar{x}|/s_n = |d_d|/s_n. \tag{31.1}$$

Values of (d_d/s_n) for given sample sizes N are shown in Chauvenet's table below (table 31.1).

Table 31.1. Values of τ_{max} for a given sample size N.

Sample size, N	τ_{max}
3	1.38
4	1.54
5	1.65
6	1.73
7	1.80
8	1.87
9	1.91
10	1.96
15	2.13
20	2.24
25	2.33
50	2.57
100	2.81
300	3.14
500	3.29
1000	3.48

Example: For $N = 9$, let the standard deviation s_n of the sample data be 1.5 units. For $N = 9$, from the above table, $\tau_{max} = d_d/s_n = 1.91$. Thus, any suspected reading (x_d) in the sample data can be checked to see whether it deviates from the mean $\bar{x}$ by $1.91 \times 1.5 = 2.865$ units. If this is the case, it may be rejected as an outlier.

The outlier is removed from the sample data and a new mean and standard deviation are calculated with a revised sample size = (n−number of outliers).

A fact one should know

Rejecting an outlier is a somewhat disputed subject, particularly when the sample size is small, because with a small sample size it is difficult to justify the assumption of a Normal distribution. Sometimes, an outlier is unambiguous; in such cases, it may indicate some deficiency in the process or point to a periodic systematic effect. Rejection is justified when one is confident about the distribution of the data and the measurement process. An outlier can also occur due to a blunder in manual data recording!

As standard deviation is very sensitive to outliers, the removal of outlier(s) from a data set improves (i.e. reduces) the standard deviation of the remaining data. The change in the mean $\bar{x}$ is generally not that large.

References

[1] ASME B89.7.3.1-2001 *Guidelines for Decision Rules: Considering Measurement Uncertainty in Determining Conformance with Specifications* (New York, NY: ASME) [Reaffirmed in 2019]
[2] ANSI ASTM E 178 2002 *Standard Practice for Dealing with Outlying Observations* (Washington, DC: ANSI)

Chapter 32

Analyzing the results

Abstract: The evaluation of uncertainty becomes a routine task once a metrologist gains knowledge and experience. The task becomes even easier if software is used. But the crux of the evaluation is the analysis of various uncertainty components and to use that analysis to further reduce expanded uncertainty.

> *A statistical analysis, properly conducted, is a delicate dissection of uncertainties, a surgery of suppositions.*
>
> —M J Moroney

> *Torture numbers, and they'll confess to anything.*
>
> —Gregg Easterbrook

The central limit theorem (CLT) is the backbone of the uncertainty evaluation process. However, it is important to check that its conditions are satisfied (see chapter 14, 'How is it that we can combine different distributions?'). One gets accustomed to the task of uncertainty evaluation over a period, but it is necessary to be able to analyze each component vis-à-vis combined uncertainty, particularly w.r.t. condition #1 of the theorem. This condition requires that all the input quantities are close to each other, which means there is no dominant component of uncertainty from the Type A or Type B methods of evaluation. For a well characterized process, the dominant component usually arises from a uniform distribution in Type B evaluation.

A fact one should discern

The famous define–measure–analyze–improve–control (DMAIC) cycle of the Six Sigma process can very well be applied after uncertainty evaluation, except for the 'define' part. Every laboratory endeavors to reduce uncertainty, because lower uncertainty is a

---

hallmark of its better capability in all respects vis-a vis other laboratories. Such evaluation is a routine task, but a pragmatic metrologist delves into an analysis of the results so as to improve and control the process in its entirety. Another quality control tool is the Pareto diagram, which is based on famous 80:20 rule: 'trivial many, vital few.' By tackling the vital few, one can achieve significantly better results!

Let us take an example of the analysis of results in the form of a case study. It is depicted to emphasize the importance of analysis and meant for illustration only.

The measurement issue: Micrometer calibration using a '0' grade slip gauge of 25 mm nominal length.

The following information is provided, and we are required to estimate uncertainty at a 95% confidence level.

- Measured values: 25.01, 25.00, 25.03, 24.99, and 24.98 mm.
- Standard reference temperature—$T_{ref} = 20\ °C$.
- Calibrated value of slip gauge—25.0001 mm ± 0.000 08 mm
- Least count of the thermometer used for temperature measurement is 1 °C.
- Thermal coefficient of expansion of the material—α: $11.5 \times 10^{-6}/°C$

(Intermediate calculated values are shown with more significant digits for illustration and to avoid cumulative rounding error.)

Type A evaluation of uncertainty
- Mean value (sample mean) $\bar{x} = 25.002$ mm
- Variance of the sample $= \sum(x_i - \bar{x})^2 /(n-1) = 0.000\ 37$ mm
- Standard deviation of the sample: $\sigma = \sqrt{variance} = 0.019\ 235$ mm $= 19.235\ \mu m$
- Standard deviation of the sample mean, $a_1 = \sigma/\sqrt{n} = 19.235/\sqrt{5} = 8.602\ \mu m$
  *(See block below) (DoF = 4)

A fact one should know

Although the SI unit of length is the meter (m), many metrologists, particularly in dimensional metrology, are prone to pronounce and write '**micron**' (shown by the symbol 'μ' and often mistyped as lowercase 'u') as a unit of estimated uncertainty, instead of '**micrometer**', which has the symbol 'μm.' This is still a widely prevalent practice, but it is not in sync with SI system of units. As per the 13th General Conference of Weights and Measures (CGPM) in 1967, the use of 'micron,' which until then had been a unit of small lengths, was abrogated, because the symbol 'μ' (micro), which had been used for that unit was adopted as a prefix (for 10^{-6}) for SI units. Hence, care needs to be taken when pronouncing as well as writing the unit 'μm' and to avoid use of 'micron' as a unit of very small dimension.

One may refer to BIPM's site for the correct use of SI units. It is a very interesting and equally enlightening read and necessary for metrologists.

Type B evaluation of uncertainty

The uncertainty due to the effect of the least count of the thermometer:

1. The uncertainty due to the thermal coefficient of expansion of the gauge block

$$= 25 \times 11.5 \times 10^{-6} \times 1$$

$$= 287.5 \times 10^{-6}\ \text{mm} = 0.2875\ \mu\text{m}.$$

Hence, standard uncertainty $b_1 = 0.2875/\sqrt{3} = 0.166\ \mu\text{m}$ (assuming a rectangular distribution).

2. The uncertainty due to the difference between the temperature of the slip gauge and that of the micrometer: this is assumed to be 0.5 °C

$$= 25 \times 11.5 \times 10^{-6} \times 0.5 = 143.75 \times 10^{-6}\ \text{mm}$$

$$= 0.14475\ \mu\text{m}.$$

Standard uncertainty $b_2 = 0.143\ 75/\sqrt{3} = 0.0829\ \mu\text{m}$ (assuming a rectangular distribution).

3. The uncertainty due to the difference in the thermal expansion coefficients of the slip gauge and the micrometer: as the materials of both are the same, this is assumed to be insignificant; hence, $b_3 = 0$.

4. The uncertainty due to the flatness of the micrometer faces: 1½ fringe (say) $\approx$ 0.5 μm. Standard uncertainty, $b_4 = 0.5/\sqrt{3} = 0.288\ \mu\text{m}$ (assuming a rectangular distribution).

5. The uncertainty due to the parallelism of the micrometer faces: 1½ fringe (say) $\approx 0.5\ \mu\text{m}$. Standard uncertainty, $b_5 = 0.5/\sqrt{3} = 0.288\ \mu\text{m}$ (assuming a rectangular distribution).

6. The uncertainty component from the calibration certificate for the slip gauge as mentioned in the calibration certificate $= 0.08\ \mu\text{m}$. Standard uncertainty $b_6 = 0.08/k = 0.08/1.96 = 0.0408\ \mu\text{m}$ (assuming a Normal distribution with 95% CL and $k = 1.96$.)

The combined uncertainty, $u_c = \sqrt{(u_A{}^2 + u_B{}^2)}$

$$= \sqrt{(a1^2 + b1^2 + b2^2 + b3^2 + b4^2 + b5^2 + b6^2)}$$

$$= \sqrt{\left\{ (8.602)^2 + (0.166)^2 + (0.0829)^2 + (0)^2 + (0.288)^2 + (0.288)^2 + (0.0408)^2 \right\}}$$

$$= 8.614\ \mu\text{m}.$$

The degrees of freedom for all Type B components are assumed to be ∞.

The effective degrees of freedom (ν_{eff}) from the Welch–Satterthwaite formula

$$= 4.02 \approx 4.$$

From Student's t-distribution, $k_p = t_p(\nu_{\text{eff}}) = 2.78$ at an approximate level of confidence of $p = 95\%$.

Hence, the expanded uncertainty $U = k_p \cdot u_c = 2.78 \times 8.614 = \pm\ 23.95\ \mu\text{m}$

Result: The measured value of the micrometer at 20 °C was 25.00 mm ± 23.95 µm at a coverage factor k of 2.78 based on the *t-distribution* and effective degrees of freedom ν_{eff} equal to 4, corresponding to a coverage probability of approximately 95%.

32.1 Analysis of the results

From the above example, the following inferences can be drawn:
1. The repeatability of 8.602 µm is the most dominant factor, giving a relative uncertainty of $8.602 \times 100/8.614 = 99.86\%$, which is as good as 100%. This means any of the following causes are possible:
   (a) The operator was unskilled/not properly trained.
   (b) He/she might have made blunders while recording data or has a personal bias.
   (c) He/she recorded data without performing actual measurements. (From a cursory look at the set of five readings, one might feel that repeatability precision is quite good!)
   (d) A larger number of readings, say ten, would probably reduce the repeatability error, which contributes nearly 100% to U_c, although the reported mean shows zero error with five readings.

Points to ponder

Point (d) above underscores the importance of uncertainty as a measure of measurement quality and also highlights the distinction between error and uncertainty. Uncertainty is a better measure of measurement quality, particularly when the error of measurements of the same measurand carried out using two different methods is zero. Zero error may lead one to believe that the measurement is perfect. Many metrologists are predisposed to think that a very close reading to the reference value implies that their measurement process (method, procedure, equipment etc.) is necessarily good. Hence, out of two (or more) methods, the one that gives the closer reading to the reference value is believed to be a better process, which may not be so.

   (e) The standard itself is worn out or damaged.
   (f) The unit being calibrated is worn out, has damaged faces or has developed some play, or has dust/dirt/oil on its faces.
2. Since repeatability is itself a Normal distribution and is the most dominant, the resulting combined distribution can be assumed to be a Normal distribution. The other factors from Type B are comparable except for the factor due to the standard's uncertainty, but in total, all are almost negligible compared to the Type A uncertainty.
3. The uncertainties of flatness and parallelism are not mentioned in the given information but taken based on past experience. This is to emphasize clause 3.4.8 of the GUM [1], which underlines the importance of process

knowledge. Although their contribution is negligible as compared to that of the Type A uncertainty, it is the largest contributor compared to the other factors of Type B. In the absence of these two factors, the combined uncertainty would have been 8.604 μm, out of which the contribution of Type A would have been 8.602 μm, meaning that repeatability would have almost entirely contributed to the total uncertainty! This is a very unlikely scenario and would warrant the immediate attention of the authority.

After investigation, problem 1(a) was identified to be the main cause. After elaborate training, the results were found to have improved, and the repeatability contribution was reduced to nearly 35% of u_c.

If the cause had been other than 1(a), improvements related to control of temperature, a thermometer of higher resolution, etc. would have been required.

Reference

[1] JCGM 100:2008 *(GUM) Evaluation of Measurement Data—Guide to the Expression of Uncertainty in Measurement* 1st edn (BIPM, IEC, IFCC, ISO, IUPAC, IUPAP, OIML— International Organization for Standardization)

IOP Publishing

A Practical Handbook on Measurement Uncertainty
FAQs and fundamentals for metrologists
Swanand Rishi

Chapter 33

The proper reporting of uncertainty

Abstract: Various standards and guides specify requirements for reporting uncertainty. The main purpose of reporting uncertainty is to provide clear and unambiguous understanding of the reported values.

> *Then there is the man who drowned crossing a stream with an average depth of six inches.*
>
> —W I E Gates

ISO/IEC 17025:2017 [1] requires that the result and its uncertainty be reported in the calibration report/certificate. In the case of a test report/certificate, as per clause 7.8.3.1 (c), where applicable, the uncertainty shall be reported under three situations. The GUM [2] advises that uncertainty should be reported with the maximum possible detail for appropriate interpretation. (In the words of the GUM, 'it is preferable to err on the side of providing too much information rather than too little' and 'Have I provided enough information in a sufficiently clear manner that my result can be updated in the future if new information or data become available?')

The following gives the gist of the important guidance in the GUM. These guidelines underline a practical approach and prudence because the whole ambit of uncertainty evaluation is based on experience, intuition, assumptions, limited knowledge, and degree of belief. Hence, giving the result to an excessive number of digits does not add any value to the result, nor does it imply better quality of measurement.

1. If the measure of uncertainty is a combined uncertainty $u_c(y)$, clause 7.2.2 provides four examples of reporting the result. However, among these, it discourages the format $Y = y \pm u_c(y)$, as this format employs the $\pm$ sign, which is conventionally used for expanded uncertainty. This format may mislead one to perceive u_c as expanded uncertainty despite providing an explanation. In addition, in spite of, and in absence of the explanation, it

doi:10.1088/978-0-7503-6462-1ch33

may convey a coverage factor of $k = 1$ corresponding to a level of confidence of approximately 68%.

2. When the measure of uncertainty is an expanded uncertainty, the GUM proposes the format $Y = y \pm k u_c(y)$ with the additional information of the level of confidence, corresponding coverage factor, and effective degrees of freedom.

3. The result (calculated mean) and uncertainty should not be reported to an excessive number of digits. In particular, the uncertainty should not be given to more than two significant digits.

4. Correlation coefficients should be given with three-digit accuracy if their absolute values are *near* unity.

Some other considerations:

A. The result of measurement taken on an *indicating* instrument should be reported to the actual resolution of indication, following the rounding guidelines while calculating the mean. For example, if the calculated mean of a pressure gauge that has a resolution up to two decimal points is 25.0677 kg m^{-2}, the mean should be reported as 25.07 kg m^{-2}. Note that intermediate calculations should not be rounded to the actual resolution, as this may result in unwarranted error in the final result.

B. In the case of artifacts or certified reference materials (CRMs), the mean should be reported to the desired degree of accuracy. One good option is to report it with the same number of digits as that used for the uncertainty after rounding (which is restricted to two significant digits). Significant digits are nonzero digits, and reporting may change depending upon the unit of the mean and the uncertainty. For example, let the result of the calibration of a resistor that has a nominal value of 1 Ω measured on an $8 - 1/2$ digit DMM be 1.000 034 82 Ω. Let the estimated expanded uncertainty be 4.516 $\mu\Omega$. The overall result could be given as

$$R = 1.000\ 034\ 82\ \Omega \pm 4.52\ \mu\Omega.$$

Here, the uncertainty is rounded to two significant decimal digits (in unit of $\mu\Omega$), matching with the resolution of the indicating instrument. Alternatively, it could be given as

$$R = (1.000\ 034\ 82 \pm 0.000\ 004\ 52)\ \Omega$$

In this form, the mean and uncertainty are expressed in same unit, and the format is as per the GUM recommendation.

It should be noted that the last form should not be reported as $R = 1.000\ 034\ 82 \pm 0.000\ 004\ 52\ \Omega$ (the mean value is without a unit!). When a common unit is used, it should always be preceded by a bracket enclosing the relevant values. Otherwise, give separate units for both, even if the units are the same. For example, the ongoing result may be written as $R = 1.000\ 034\ 82\ \Omega \pm 0.000\ 004\ 52\ \Omega$.

It is worth remembering that uncertainty is to be given at *up to* two significant digits; not always *to* two significant digits.

A fact one should know

The number of significant figures (digits) roughly corresponds to the precision of measurement, not accuracy. This means that even if the resolution of a standard indicating instrument is greater, if the estimated uncertainty is comparatively large (and this is quite common), the mean value should be truncated in line with the significant digits of the uncertainty.

The following are rules of thumb for the identification of significant digits.
1. All nonzero digits are significant.
2. Zero(s) between two significant digits are significant.
3. Trailing zeros or the final zero after the decimal point are significant.

Most importantly, one should be honest when reporting a measurement. A digit is significant if it (the digit) is required to express the numerical value to the required precision. The reported result should not appear to be more accurate and precise than the capabilities of the instruments and the process employed for the measurement.

C. The expanded uncertainty may also be reported as a percentage relative to the mean value. For example, in case of the above example for the resistor, the relative uncertainty is

$$= (0.000\,004\,52/1.000\,034\,82) \times 100$$

$$= 0.000\,45\%.$$

The percentage form is easy for comparison with other results.

D. In the case of the evaluation of uncertainty of a measurand given as a function of various input quantities that have different units, it is easier to estimate the individual uncertainty contributions by converting them to relative standard uncertainties before combining their uncertainty contributions. (This is a small digression from the main subject matter of this topic, but it is important as far as *intermediate* calculations in such cases are concerned. Notwithstanding, the main theme is the need to express the individual contributions in terms of relative uncertainty.)

This is a common scenario in analytical and mechanical measurements. For example, a solution of sodium hydroxide (NaOH) is to be standardized against the titrimetric standard potassium hydrogen phthalate (KHP). In this method, the equation for the 'concentration' of a NaOH solution is given by

$$c_{NaOH} = \frac{m_{KHP} \cdot P_{KHP}}{M_{KHP} \cdot V_T} \times 1000 \text{ mol l}^{-1}. \qquad (33.1)$$

We shall not delve into the details of each factor but consider just one factor for illustration. From the equation, one contributory factor in the method is M_{KHP}, the molar mass of KHP. Its value is 204.22 g mol^{-1} with a standard uncertainty of 0.0064 g mol^{-1}, and its relative standard uncertainty is

$$
\begin{aligned}
&= u(M_{KHP})/M_{KHP} \\
&= 0.0064/204.22 \\
&= 0.000\ 031 (\text{obviously a unit}-\text{free value}).
\end{aligned}
\tag{33.2}
$$

Other uncertainty contributions should be converted similarly and then combined by the RSS method as usual. The combined uncertainty u_c of c_{NaOH} is obtained by multiplying the RSS value by the value of c_{NaOH} calculated using the above equation.

A fact one should know

It should be noted that with this method, the sensitivity coefficients do not come into the picture, even though the measurand is given by an equation involving contributory factors that have different units. (See chapter 25 'What is the significance of the sensitivity coefficient?') Sensitivity coefficients would enter the picture if the partial derivatives of the measurand were taken w.r.t. each contributory factor. However, that method is more cumbersome and needs sound command of calculus.

References

[1] ISO/IEC 17025:2017 *General Requirements for the Competence of Testing and Calibration Laboratories* (Geneva: ISO)

[2] JCGM 100:2008 *(GUM) Evaluation of Measurement Data—Guide to the Expression of Uncertainty in Measurement* 1st edn (BIPM, IEC, IFCC, ISO, IUPAC, IUPAP, OIML—International Organization for Standardization)

Chapter 34

Alternative approaches in uncertainty evaluation

Abstract: JCGM 100:2008 (GUM) or ISO/IEC Guide 98-3 [1, 2] is the most popular and widely followed guide for evaluation of uncertainty based on a mathematical modeling approach. However, it should be noted that, apart from these, there are other approaches for the evaluation of uncertainty.

> *If you do not change direction, you may end up where you are heading.*
> —Lao Tzu

> *Taking a model too seriously is really just another way of not taking it seriously at all.*
> —Andrew Gelman

It should be noted that the GUM's philosophy is largely based on a mathematical modeling approach and is widely perceived to be the only way to evaluate uncertainty. The GUM itself states this (to express a mathematical relationship between a measurand and the input quantities) as a first step in the summary of procedure in clause 8. This perception is further substantiated by the multitude of papers and national/regional guidelines which have followed this practice and thus inculcated that impression. The GUM's approach is, by and large, applicable for a particular measurement result and is well founded in physical metrology. However, in analytical metrology, the GUM approach has limitations, since in analytical metrology there is an emphasis on the effect of the test procedure on the test objects and also the uncertainties of method, extrapolation, and bias are important. Clause F.2.5 of GUM takes cognizance of this fact, further acknowledging that uncertainty of method is most difficult to assess and could be dominant in certain cases. It suggests methods that are appropriate for analytical and quantitative tests. This chapter outlines some methods that can be used when the GUM approach is either unsuitable or inappropriate (for a variety of reasons, including applications) or calls for method validation. The GUM itself, in clause G.1.5, allows the employment of alternate analytical and numerical methods, when the conditions for its use are not met. These approaches, however, broadly follow the principles and philosophy of the GUM.

Among these alternatives:

(1) The Monte Carlo method is suitable when the conditions required for GUM's implementation are not fulfilled.

(2) The interlaboratory method and the proficiency testing method are appropriate for analytical (chemical) and quantitative testing due to the inherent requirement for method validation. These methods fall under the interlaboratory approach, which calls for collaborative evaluation by different laboratories. Since this approach focuses on the complete method, it is also called a 'top down' approach. (The GUM approach is called a 'bottom up' approach due to its emphasis on individual input quantities.) This is particularly appropriate if the major contributions to uncertainty cannot be readily modeled in terms of measurable influence quantities and where many laboratories use essentially identical test methods and equipment. The result of the performance of the procedure on a given test object is the focus of these approaches used in quantitative testing.

(3) Measurement system analysis is mostly used in the manufacturing industry by quality control personnel to establish the total variation (which comprises process and measurement variation) and to evaluate the contribution of measurement variation. Process variation is assumed to be entirely due to random effects under a statistically controlled process. Measurement variation should obviously be negligible.

Approaches based on (2) and (3) are termed empirical approaches, as they involve experiential data obtained under varied conditions.

(A) **The Monte Carlo method (MCM):** The BIPM and other metrological bodies have issued further guidance in JCGM 101:2008 [3]: 'Evaluation of measurement data—Supplement 1 to the 'Guide to the expression of uncertainty in measurement'—Propagation of distributions using a Monte Carlo Method.' This method is suitable when the conditions required for the implementation of the GUM are not fulfilled and particularly where: (a) the model is not differentiable, (b) the model is strongly nonlinear, or (c) the distributions are strongly non-normal.

Under the above conditions, the probability distribution of the measurand can greatly deviate from a Normal distribution, thereby substantially underestimating the coverage factor $k = 2$ at a 95% level of confidence. (When the above conditions are not encountered, linear regression can be used, allowing standard uncertainties to be safely propagated.)

A fact one should know

It should be noted that the GUM uncertainty framework is based on the central limit theorem (CLT), which employs distributions of *estimates* x_i of input quantities (rather than input quantities X_i) and produces an *estimate y* of the output quantity (rather than the output quantity Y). Thus, the results produced are approximations and may turn out to be far from reality if the conditions of the CLT are not met. By employing distributions of quantities X_i and Y, rather than their *estimate*, the MCM overcomes these limitations.

This problem can be circumvented if, instead of the standard uncertainties, the probability distributions attributed to the input quantities are combined (propagated). This is done in the MCM by attributing a suitable distribution (usually normal, rectangular, or triangular) to each input quantity. A random value is generated for each, and a value of the target quantity is calculated from the set of sampled input data, which is required to be large (usually numbering a few thousand values). The mean and the standard deviation of this random sample represent the estimates of the value of the target quantity and its standard uncertainty. The MCM additionally provides estimates of expectation, variance, and coverage intervals of the output quantity, even for higher moments. It also provides the distribution of values attributed to the target quantity based upon the available information about the input quantities. The outcome of the MCM depends upon the number of trials, (the more the better). Thus, the simulated distribution provides a more realistic confidence interval under the above conditions. The MCM is also for validation of results obtained using the GUM framework.

Some prominent features of the MCM (as compared to the GUM framework) are:

1. PDFs are assigned to the quantities themselves rather than to standard deviations of estimates of those quantities.
2. Classification into Type A and Type B methods of evaluation is not necessary.
3. Sensitivity coefficients or partial derivatives of the function are not required.
4. The output quantity can be defined by a non-Gaussian distribution and it is usually asymmetrical. The coverage interval thus is not centered on the estimate of output quantity and needs to be carefully defined.
5. Less effort is required for evaluation, particularly with higher-order non-linear models.
6. It provides better results.
7. A coverage factor is not required to define the coverage interval.

The following example illustrates the use of the MCM for the Brinell Hardness test. This test is conducted by applying a force F to a sphere made of a hard material (usually steel) of diameter D, which is placed over the surface of the sample under test, as shown in figure 34.1.

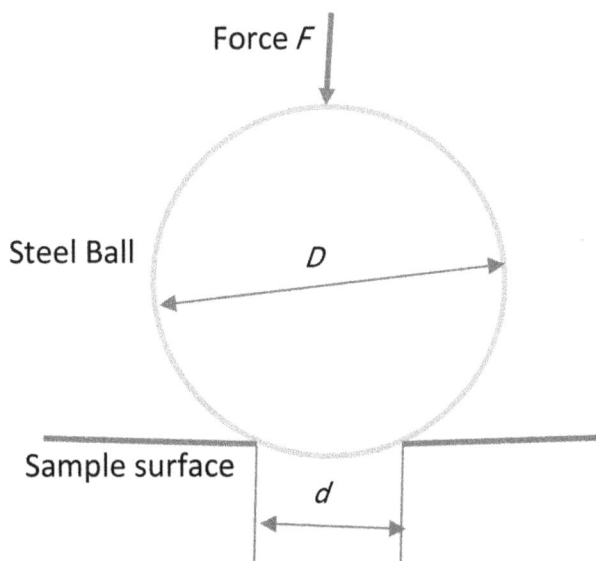

Figure 34.1. The Brinell Hardness test.

The sphere leaves an indentation mark of diameter d on the sample due to the applied force. The diameter of this mark (called the indentation diameter) is inversely proportional to the hardness of the material of the sample.

The equation for Brinell Hardness B is given as

$$B = 0.204 \times F/\pi \times D\left\{D - \sqrt{(D^2 - d^2)}\right\}$$

where F is the applied load (N), D is the ball diameter (mm), and d is the indentation diameter (mm).

Let the force applied $= 3000$ kgf (i.e. 29 400 N at $g = 9.8$ m s^{-2}), for which the uncertainty of force indication from the certificate is $+/-1\% = +/-294$ N.

A 10 mm steel ball was used for the test, giving diameters of indentation of: 3.00, 3.10, 2.90, 3.05, 2.95 mm, giving a mean of 3.00 mm.

So, the standard deviation of indentation for five readings $= 0.079$ mm.

Let the expanded uncertainty (standard deviation) of the diameter of the ball (from the certificate) $= +/-0.01$ mm. At an approximate level of confidence of 95%, i.e. at $k = 2$, the standard uncertainty $= 0.01/2 = 0.005$ mm).

The results of uncertainty evaluation as per the GUM (using the Gaussian linear approximation method) are:
- B (mean) = 414 HBS
- combined uncertainty = 23 N mm^{-2}

The results of uncertainty evaluation as per MCM for sample of size 200 000 are:
- B (mean) = 415 HBS
- combined uncertainty (SD) = 23 N mm^{-2}
- median = 415 HBS
- median absolute deviation (MAD) = 23 N mm^{-2}

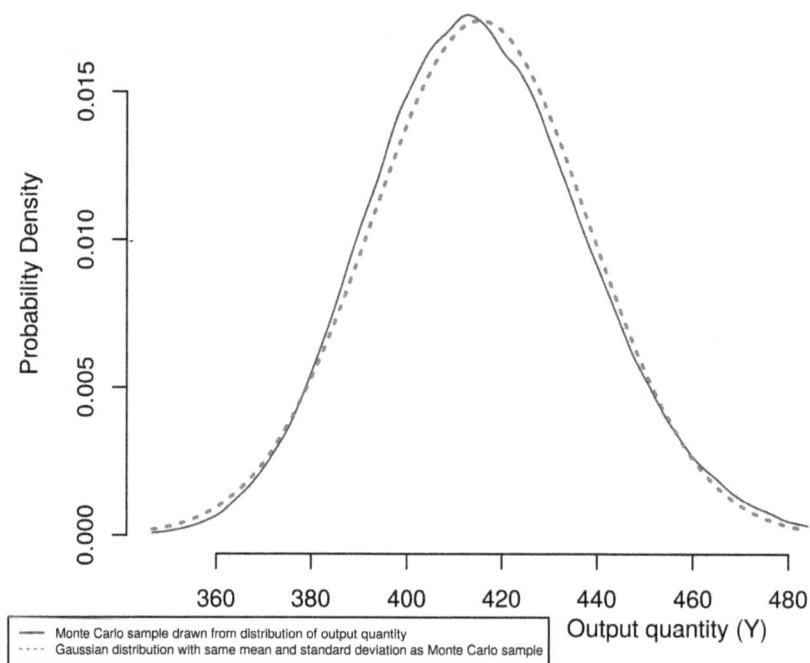

Figure 34.2. PDFs (distributions) for linear inputs produced by the GUM method. Generated using the NUM [4].

From the above example and figure 34.2, it can be seen that both approaches give nearly the same results, implying that there is a fair amount of linearity in the input parameters.

If there is large nonlinearity, the GUM approach may give substantially different but more faithful results.

For example, assuming that the standard uncertainty of indentation for five readings is 0.35 mm (instead of the figure of 0.079 mm obtained earlier), indicating a large amount of nonlinearity and assuming that the other inputs remain same, the results of uncertainty evaluation as per the GUM (Gaussian linear approximation method) are:

B (mean) $= 410$ HBS
combined uncertainty $= 99$ N mm^{-2}

The results of uncertainty evaluation as per MCM for the same sample of size 200 000 are:

B (mean) $= 430$ HBS
combined uncertainty (SD) $= 110$ N mm^{-2}
median $= 410$ HBS
MAD $= 98$ N mm^{-2}

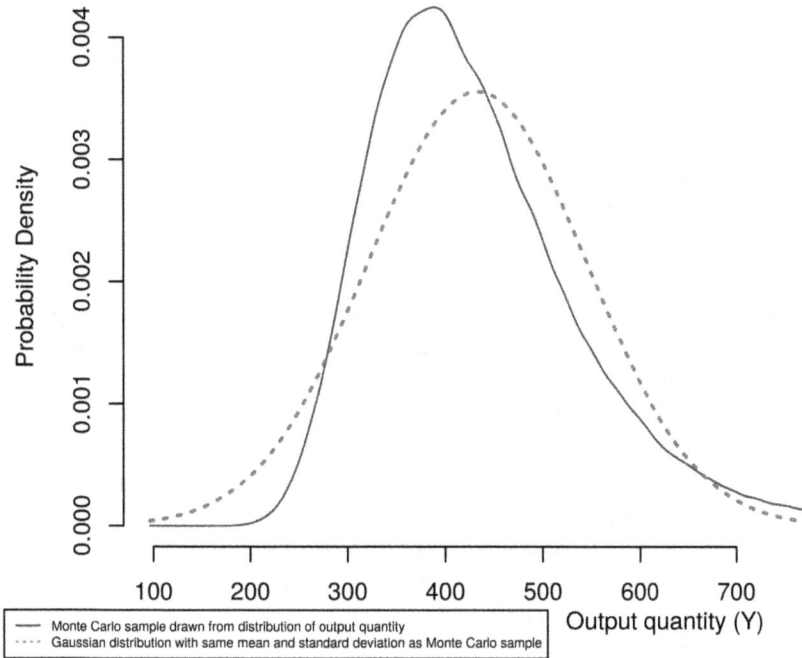

Figure 34.3. PDFs (distributions) for nonlinear inputs produced by the GUM method.

Figure 34.3 shows that the resultant output distribution is highly skewed toward the left. This means that the GUM result is misleading in the case of highly nonlinear input(s) and if it is assumed that the output distribution is a t-distribution.

It may also be noticed that in cases of high nonlinearity, the median (the value in a data set that divides data equally on either side of it when the data is arranged in ascending or descending order) is a more correct representation of central tendency than the mean value.

Note: *For both examples of Brinell Hardness (linear and nonlinear inputs), the NIST Uncertainty machine (NUM) was used to evaluate uncertainty by the MCM. It also gave results produced using the GUM methodology. It is a web-based software application that evaluates the measurement uncertainty associated with an output quantity defined by a measurement model of the form> $y = f(x_0,...,x_n)$.*

The NUM has been used here just to illustrate the GUM and the MCM within its framework. For more details, the reader may refer to the NUM [4].

(B) **Empirical approaches:** in quantitative testing, particularly in chemical and food laboratories, certain well-founded test procedures are involved, but evaluation of the uncertainty of the test procedure calls for more practical data from similar experiments done by other laboratories. Thus, it is required to factor in as many sources of uncertainty as possible for a meaningful evaluation of uncertainty. In these fields, precision and bias data from intralaboratory and interlaboratory tests are typically used for that purpose. In this process, however, the basic tenets of the GUM philosophy are adhered to.

The methods that are considered to be more appropriate in these tests are termed empirical approaches, which include

    A. The single (or intra) laboratory validation approach
    B. The interlaboratory validation approach
    C. The proficiency testing (PT) validation approach

(A) **The single laboratory validation approach:** this falls under the intralaboratory approach. (The modeling approach is also a single (or intra) laboratory validation approach)

The protocol used for intralaboratory validation consists of:

- an investigation of precision,
- an investigation of trueness (trueness is a recommended term and is the opposite of bias, although the GUM does not use it), and
- a correction of bias, if significant (uncertainty of correction of bias, even if the bias is insignificant and hence neglected, shall however, be considered).

The uncertainty under this approach is given by

$$u = \sqrt{\left(s^2 + b^2\right)}, \tag{34.1}$$

where s is the standard deviation of the precision of measurement. This is obtained by regular measurements. If it is obtained using pooled standard variance, s is given by

$$s^2 = \frac{(n_1 - 1)s_1^2 + (n_2 - 1)s_2^2}{n_1 + n_2 - 2}, \tag{34.2}$$

where n_1 and n_2 are the numbers of samples from which s_1 and s_2 are obtained. b is the uncertainty of bias, which is given by

$$b = \sqrt{(\Delta^2 + u_{\text{ref}}^2 + s_b^2/n)}. \tag{34.3}$$

In equation (34.3), Δ is the mean deviation of the replicate measurements, u_{ref} is the uncertainty of the reference value, s_b is the standard deviation of the precision of the mean value of the replicate measurements made in the bias investigation, and n is the number of readings.

(B) **The interlaboratory approach:** in this approach, well-established standard procedures are followed. 'Generally the test conditions as specified in ISO 21748: Guide to the use of repeatability, reproducibility, and trueness estimates in measurement uncertainty estimation' are followed, and the reproducibility standard deviation, s_R, assigned to a standard test procedure is used as an estimate of the measurement uncertainty. The repeatability standard deviation s_R is not suitable for this purpose (as it does not include important contributions), but the laboratory has to check that its value is compatible with interlaboratory data. Further, s_R normally covers laboratory bias; any bias due to method is usually insignificant and is neglected.

Thus uncertainty in this approach is given by

$$u = s_R, \tag{34.4}$$

where s_R is the reproducibility standard deviation.

(Note: if there is large departure from the test conditions, procedures, or objects, the uncertainties of these other factors have to be included in the above uncertainty using the root sum square (RSS) method.)

Thus, in the case of large departures from the test conditions, the uncertainty

$$u = \sqrt{\left(s_R^2 + \sum u_{other}^2\right)}. \tag{34.5}$$

(C) **The proficiency testing (PT) validation approach:** in this approach, the standard deviation of precision s is the intralaboratory reproducibility standard deviation S_{Rw}, and the bias b is taken from the data obtained by participating in the PT program.

Thus, the uncertainty in this approach is given by

$$u = \sqrt{(s^2 + b^2)} \tag{34.6}$$

(Note that this equation is the same as the single laboratory approach, equation (34.1)).

The uncertainty of the bias is given by

$$b = \sqrt{(\Delta^2 + u_{ass}^2 + s_{PT}^2/n)}, \tag{34.7}$$

where Δ is the deviation of the laboratory's result (or the mean deviation of the replicate measurements) from the assigned value of the PT sample, u_{ass} is the assigned uncertainty of the PT sample, and s_{PT} is the standard deviation of the precision for the PT sample, with n readings.

In the PT approach, the two basic components of uncertainty are obtained from different investigations: precision is obtained by an in-house validation method, while bias is estimated from PT results.

The data obtained from PT programs are much debated as far as quantitative tests are concerned due to lack of consensus on the procedure as well as the validation of the data thus obtained. However, PT studies are gaining increasing acceptance and their data is being used for uncertainty evaluations of ongoing measurements as well as the verification of uncertainty estimates produced by other methods.

The data obtained from a laboratory's participation in PT can be a good basis for uncertainty estimates, subject to the following conditions:

- The test items should reasonably represent the routine test items (for example, the type of material and range of values of the measurand).
- The assigned values have an appropriate uncertainty.
- The number of PT rounds is appropriate. A minimum of six different trials over an appropriate period is recommended

Interested metrologists may refer to:

- ISO/IEC 17043:2010—Conformity assessment—General requirements for proficiency testing [5]
- ISO 13528: 2022—Statistical methods for use in proficiency testing by Interlaboratory comparisons [6]
- ISO Guide 43-1:1997—Proficiency Testing by Interlaboratory Comparisons—Part 1: Development and Operation of Proficiency Testing Schemes [7]
- ISO/IEC Guide 43-2:1997—Proficiency testing by Interlaboratory comparisons—Part 2: Selection and use of proficiency testing schemes by laboratory accreditation bodies [8]

(D) **Measurement system analysis (MSA):** This method is popular in quality control departments in the manufacturing industry, where process variation is a critical parameter to be measured and controlled. It is widely used for assigning errors to all elements of the measurement system—operator, equipment, test object, environment, and procedure. Various contributions that are considered include:

- repeatability standard deviation,
- intermediate-precision standard deviation,
- reproducibility standard deviation,
- bias estimates and,
- data from method validation studies and quality control.

This method is also called R&R (repeatability and reproducibility) study, in which the ratio of measurement variation to process variation (calculated in terms of standard deviation) called gauge R&R (GRR) is established. Repeatability is also known as equipment variation and reproducibility as appraiser variation. A smaller GRR is naturally desired. General acceptance norms for GRR are:

a. If the GRR is <10%: accept.
b. If the GRR is between 10% and 30%: acceptable, depending upon criticality of application, repair, and other cost implications. Thus, it is a techno–commercial decision.
c. If the GRR is >30%: reject.

In these studies, ten objects are generally measured by three appraisers with a set of three readings each. Bias and linearity (bias over the operating range) estimates are also performed in MSA.

References

[1] ISO/IEC Guide 98-3 2008 *Uncertainty of Measurement—Part 3: Guide to the Expression of Uncertainty in Measurement* 1st edn (GUM:1995)

[2] JCGM 100:2008 *(GUM) Evaluation of Measurement Data—Guide to the Expression of Uncertainty in Measurement* 1st edn (BIPM, IEC, IFCC, ISO, IUPAC, IUPAP, OIML—International Organization for Standardization)

[3] JCGM 101:2008 *Evaluation of measurement data — Supplement 1 to the "Guide to the expression of uncertainty in measurement" — Propagation of distributions using a Monte Carlo method* (Sèvres: BIPM)

[4] NIST 2024 NIST Uncertainty Machine https://uncertainty.nist.gov/

[5] ISO/IEC 17043:2010 2010 *Conformity assessment—General requirements for proficiency testing* (Geneva: ISO)

[6] ISO 13528:2022 *Statistical methods for use in proficiency testing by Interlaboratory comparisons* (Geneva: ISO)

[7] ISO Guide 43-1:1997 *Proficiency Testing by Interlaboratory Comparisons—Part 1: Development and Operation of Proficiency Testing Schemes* (Geneva: ISO)

[8] ISO/IEC Guide 43-2:1997 *Proficiency testing by Interlaboratory comparisons—Part 2: Selection and use of proficiency testing schemes by laboratory accreditation bodies* (Geneva: ISO)

IOP Publishing

A Practical Handbook on Measurement Uncertainty
FAQs and fundamentals for metrologists
Swanand Rishi

Chapter 35

Some important notes in the GUM

Abstract: All standards and guides include many notes under the various clauses that are meant to explain or interpret those clauses. These notes form an important part of the document but are usually overlooked by readers.

> *Knowledge is an unending adventure at the edge of uncertainty.*
> —Jacob Bronowski

The majority of the concepts presented in this book are based on the GUM [1] philosophy, although it also refers to other resources. This topic is intended to familiarize the reader with some important aspects that have been elaborated in the GUM in the form of notes and examples, although most of them have been elaborated in other topics in this book. The direct presentation of notes from the GUM will have more impact and hopefully prompt the reader to browse through the GUM and set him/her mulling over the finer aspects of uncertainty evaluation. It is important that the notes should be read in conjunction with the relevant subject and clauses in the GUM and not in isolation.

The notes are presented in the form of tables in three categories—general, cautionary, and suggestive. The notes have been given as they appear in the GUM so that the interested readers can refer to the references (clauses, equations, examples, etc.) mentioned therein.

(The page numbers in column 1 of tables 35.1–35.3. refer to the version of the GUM available on the BIPM website. The reader is advised to preferably refer to the clause number.)

Table 35.1. General notes/comments.

Page no.	Clause no.	Note
1	1.4	There may be situations in which the concept of uncertainty of measurement is believed not to be fully applicable, such as when the precision of a test method is determined (see Reference [5], for example).
5	3.2.4	Note 1: Often, measuring instruments and systems are adjusted or calibrated using measurement standards and reference materials to eliminate systematic effects; however, the uncertainties associated with these standards and materials must still be taken into account.
7	3.3.7	The coverage factor k is always to be stated, so that the standard uncertainty of the measured quantity can be recovered for use in calculating the combined standard uncertainty of other measurement results that may depend on that quantity.
10	4.2.3	Note 2: Although the variance $s^2(q)$ is the more fundamental quantity, the standard deviation $s(q)$ is more convenient in practice because it has the same dimension as q and a more easily comprehended value than that of the variance.
21	5.2.2	Note 1: For the very special case where *all* of the input estimates are correlated with correlation coefficients $r(x_i, x_j) = +1$, equation (16) reduces to $u_c^2(y) = \left[\sum_{i=1}^{N} c_i u((x_i)) \right]^2 = \left[\sum_{i=1}^{N} \frac{\partial f}{\partial x_i} u((x_i)) \right]^2$. The combined standard uncertainty $u_c(y)$ is thus simply a *linear sum* of terms representing the variation of the output estimate y generated by the standard uncertainty of each input estimate x_i (see 5.1.3). [This linear sum should not be confused with the general law of error propagation although it has a similar form; standard uncertainties are not errors (see E.3.2).]
32	B.2.3	Note 3: The indefinite article 'a,' rather than the definite article 'the,' is used in conjunction with 'true value' because there may be many values consistent with the definition of a given particular quantity.
34	B.2.11	Note 2: A complete statement of the result of a measurement includes information about the uncertainty of measurement.
36	B.2.17	Note 3: 'Experimental standard deviation of the mean' is sometimes incorrectly called **standard error of the mean**.
36	B.2.19	Note 1: Since a true value cannot be determined, in practice a conventional true value is used
47	C.3.7	If two random variables are independent, their covariance and correlation coefficient are zero, but the converse is not necessarily true.
50	D.3.4	Note 2: Although a measurand should be defined in sufficient detail that any uncertainty arising from its incomplete definition is negligible in comparison with the required accuracy of the measurement, it must be recognized that this may not always be practicable. The definition may, for example, be incomplete because it does not specify parameters that may have been assumed, unjustifiably, to have negligible effect; or it may imply conditions that can never be fully met and whose imperfect

		realization is difficult to take into account. For instance, in the example of D.1.2, the velocity of sound implies infinite plane waves of vanishingly small amplitude. To the extent that the measurement does not meet these conditions, diffraction and nonlinear effects need to be considered.
56	E.3.2	Equation (E.3) also applies to the propagation of multiples of standard deviations.... However, it does not apply to the propagation of confidence intervals.
56	E.3.3	The requirement of normality when propagating confidence intervals using equation (E.3) may be one of the reasons for the historic separation of the components of uncertainty derived from repeated observations, which were assumed to be normally distributed, from those that were evaluated simply as upper and lower bounds.
59	E.5.2	It is assumed that probability is viewed as a measure of the degree of belief that an event will occur, implying that a systematic error may be treated in the same way as a random error and that ε_i represents either kind.
70	G.1.3 Table G.1	By contrast, if z is described by a rectangular probability distribution with expectation μ_z and standard deviation $\sigma = a/\sqrt{3}$, where a is the half-width of the distribution, the level of confidence p is 57.74% for $k_p = 1$; 95% for $k_p = 1.65$; 99% for $k_p = 1.71$; and 100% for $k_p = \sqrt{3} \approx 1.73$; the rectangular distribution is 'narrower' than the Normal distribution in the sense that it is of finite extent and has no 'tails.'
71	G.2.2	EXAMPLE The rectangular distribution (see 4.3.7 and 4.4.5) is an extreme example of a non-Normal distribution, but the convolution of even as few as *three* such distributions of equal width is approximately normal. If the half-width of each of the three rectangular distributions is a so that the variance of each is $a^2/3$, the variance of the convolved distribution is $\sigma^2 = a^2$. The 95% and 99% intervals of the convolved distribution are defined by 1937σ and 2379σ, respectively, while the corresponding intervals for a Normal distribution with the same standard deviation σ are defined by 1960σ and 2576σ (see table G.1) [10].
71	G.2.2	Note 1: For every interval with a level of confidence p greater than about 91.7%, the value of k_p for a Normal distribution is larger than the corresponding value for the distribution resulting from the convolution of any number and size of rectangular distributions.
75	G.5.2	Note 2: For a Normal distribution, the coverage factor $k = \sqrt{3} \approx 1732$ provides an interval with a level of confidence $p = 91\ 673\%$.... This value of p is robust in the sense that it is, in comparison with that of any other value, optimally independent of small deviations of the input quantities from normality.

Table 35.2. Cautionary notes.

Page no.	Clause no.	Note
5	3.2.2	Note 2: In this *Guide*, great care is taken to distinguish between the terms 'error' and 'uncertainty.' They are not synonyms, but represent completely different concepts; they should not be confused with one another or misused.
5	3.2.3	The uncertainty of a correction applied to a measurement result to compensate for a systematic effect is *not* the systematic error, often termed bias, in the measurement result due to the effect as it is sometimes called. It is instead a measure of the *uncertainty* of the result due to incomplete knowledge of the required value of the correction. The error arising from imperfect compensation of a systematic effect cannot be exactly known. The terms 'error' and 'uncertainty' should be used properly and care taken to distinguish between them.
6	3.3.1	The result of a measurement (after correction) can unknowably be very close to the value of the measurand (and hence have a negligible error) even though it may have a large uncertainty. Thus the uncertainty of the result of a measurement should not be confused with the remaining unknown error.
6	3.3.3	In some publications, uncertainty components are categorized as 'random' and 'systematic' and are associated with errors arising from random effects and known systematic effects, respectively. Such categorization of components of uncertainty can be ambiguous when generally applied. For example, a 'random' component of uncertainty in one measurement may become a 'systematic' component of uncertainty in another measurement in which the result of the first measurement is used as an input datum. Categorizing the *methods* of evaluating uncertainty components rather than the *components* themselves avoids such ambiguity. At the same time, it does not preclude collecting individual components that have been evaluated by the two different methods into designated groups to be used for a particular purpose (see 3.4.3).
14	4.3.10	It is important not to 'double-count' uncertainty components. If a component of uncertainty arising from a particular effect is obtained from a Type B evaluation, it should be included as an independent component of uncertainty in the calculation of the combined standard uncertainty of the measurement result only to the extent that the effect does not contribute to the observed variability of the observations. This is because the uncertainty due to that portion of the effect that contributes to the observed variability is already included in the component of uncertainty obtained from the statistical analysis of the observations.
24	6.3.1	Occasionally, one may find that a known correction b for a systematic effect has not been applied to the reported result of a measurement, but instead an attempt is made to take the effect into account by enlarging the 'uncertainty' assigned to the result. This should be avoided; only in very special circumstances should corrections for known significant systematic effects not be applied to the result of a measurement (see F.2.4.5 for a specific case and how to treat it). Evaluating the uncertainty of a measurement result should not be confused with assigning a safety limit to some quantity.

36	B.2.19	Note 2: When it is necessary to distinguish 'error' from 'relative error,' the former is sometimes called **absolute error of measurement**. This should not be confused with **absolute value of error**, which is the modulus of the error.
50	D.3.4	Note 3: Inadequate specification of the measurand can lead to discrepancies between the results of measurements of ostensibly the same quantity carried out in different laboratories.
62	F.1.2.1	Note 2: Different experiments may not be independent if, for example, the same instrument is used in each (see F.1.2.3).

Table 35.3. Suggestive notes.

Page no.	Clause no.	Note
7	3.3.7	The coverage factor k is always to be stated, so that the standard uncertainty of the measured quantity can be recovered for use in calculating the combined standard uncertainty of other measurement results that may depend on that quantity.
11	4.2.8	At lower levels of the calibration chain, where reference standards are often assumed to be exactly known because they have been calibrated by a national or primary standards laboratory, the uncertainty of a calibration result may be a single Type A standard uncertainty evaluated from the pooled experimental standard deviation that characterizes the measurement.
42	C.2.19	Note 1: The term 'mean' is used generally when referring to a population parameter and the term 'average' when referring to the result of a calculation on the data obtained in a sample.
73	G.4.1	Note 1: If the value of ν_{eff} obtained from equation (G.2b) is not an integer, which will usually be the case in practice, the corresponding value of t_p may be found from table G.2 by interpolation or by truncating ν_{eff} to the next lower integer.

I would like to invoke clause 3.4.8 of GUM, which sums up the very purpose of this book:

Although this Guide provides a framework for assessing uncertainty, it cannot substitute for critical thinking, intellectual honesty and professional skill. The evaluation of uncertainty is neither a routine task nor a purely mathematical one; it depends on detailed knowledge of the nature of the measurand and of the measurement. The quality and utility of the uncertainty quoted for the result of a measurement therefore ultimately depend on the understanding, critical analysis, and integrity of those who contribute to the assignment of its value.

References and further reading

[1] JCGM 100:2008 *(GUM) Evaluation of Measurement Data—Guide to the Expression of Uncertainty in Measurement* 1st edn (BIPM, IEC, IFCC, ISO, IUPAC, IUPAP, OIML—International Organization for Standardization)

[2] ISO/IEC Guide 98-3 2008 *Uncertainty of Measurement—Part 3: Guide to The Expression of Uncertainty in Measurement (GUM:1995)* 1st edn (Geneva: ISO)

Further references

[1] NPL 2001 *A Beginner's Guide to Uncertainty of Measurement* Guide No. 11, NPL UK, Issue 2

[2] ASME B89.7.3.3-2002 *Guidelines for Assessing the Reliability of Dimensional Measurement Uncertainty Statements* (New York: ASME)

[3] EA-4/02 2013 *Evaluation of the Uncertainty of Measurement in Calibration, European cooperation for accreditation, Rev. 01, 2013* (Paris: EA)

[4] EURACHEM 2011 *Terminology in Analytical Measurement: Introduction to VIM 3* 1st edn (Gembloux: EURACHEM)

[5] EURACHEM/CITAC Guide CG 4 2012 *Quantifying uncertainty in analytical measurement* 3rd edn (Gembloux: EURACHEM)

[6] EURACHEM/CITAC Guide 2021 *Use of Uncertainty Information in Compliance Assessment* 2nd edn (Gembloux: EURACHEM)

[7] EURACHEM/CITAC Guide 2019 *Measurement Uncertainty Arising from Sampling: A Guide to Methods and Approaches* 2nd edn (Gembloux: Euratech)

[8] EUROLAB 2002 *Measurement Uncertainty in Testing* (Paris: EUROLAB) EUROLAB Technical Report 1/2002

[9] EUROLAB Technical Report 1/2006 *Guide to the Evaluation of Measurement Uncertainty for Quantitative Test Results*

[10] G104—A2LA 2019 *Guide for Estimation of Measurement Uncertainty in Testing* (Frederick, MD: A2LA)

[11] Rishi S 2012 Guard-Banding Methods-An Overview *AdMet* **2012** UM 001

[12] ISO14253–1:1998 *Geometrical Product Specifications (GPS)—Inspection by Measurement of Work Pieces and Measuring Equipment—Part 1: Decision Rules for Proving Conformance or Non-Conformance with Specifications* (Geneva: ISO) [Revised 2017]

[13] ISO 14253–2:1999 *International Standard Geometrical Product Specifications (GPS)— Inspection by Measurement of Workpieces and Measuring Instruments—Part 2: Guide to the Estimation of Uncertainty in GPS Measurement in Calibration of Measuring Equipment and in Product Verification* (Geneva: ISO) [Revised 2011]

[14] ISO 14253–3:2002 International Standard 2002 *Geometrical Product Specifications (GPS)— Inspection by Measurement of Workpieces and Measuring Instruments—Part 3: Guidelines to Achieving Agreements on Measurement Uncertainty Statements* [Revised 2011] *Geometrical Product Specifications (GPS)—Inspection by Measurement of Workpieces and Measuring*

Instruments—Part 3: Guidelines to Achieving Agreements on Measurement Uncertainty Statements (Geneva: ISO) [Revised 2011]

[15] ISO 3534–2:2006 *Statistics-Vocabulary and Symbols-Part 2: Applied Statistics* (Geneva: ISO) [confirmed in 2014]

[16] ISO 5725:1994 *Accuracy (Trueness and Precision) of Measurement Methods and Results* (Geneva: ISO)
   a. Part 1:2023—General Principles and Definitions
   b. Part 2:2019—Basic Method for the Determination of the Repeatability and Reproducibility of a Standard Measurement Method
   c. Part 3:2023—Intermediate Measures of the Precision of a Standard Measurement Method
   d. Part 4:2020—Basic Methods for the Determination of the Trueness of a Standard Measurement Method
   e. Part 5:1998 [confirmed 2018]—Alternative Methods for the Determination of the Precision of a Standard Measurement Method
   f. Part 6:1994 [confirmed 2022]—Use in Practice of Accuracy Values

[17] ISO 21748:2017 *Guidance for the Use of Repeatability, Reproducibility and Trueness Estimates in Measurement Uncertainty Evaluation* (Geneva: ISO) [confirmed in 2022]

[18] ISO/IEC 17043:2010 *Conformity Assessment—General Requirements for Competence of Proficiency Testing Providers* (Geneva: ISO) [Revised 2023]

[19] ISO/IEC Guide 33:1989 *Use of Certified Reference Materials* (Geneva: ISO) [Revised 2015]

[20] ISO/IEC Guide 98-3 2011 *Uncertainty of Measurement—Part 3: Guide to the Expression of Uncertainty in Measurement (GUM:1995) Supplement 2: Extension to any Number of Output Quantities* 1st edn (Geneva: ISO)

[21] ISO/IEC GUIDE 98-4:2012 *Uncertainty of Measurement—Part 4: Role of Measurement Uncertainty in Conformity Assessment* (Geneva: ISO)

[22] JCGM 104:2009 *Evaluation of Measurement Data—An Introduction to the 'Guide to the Expression of Uncertainty in Measurement' and Related Topics* 1st edn (Sèvres: BIPM)

[23] Birch K 2003 *Estimating Uncertainties in Testing* (Guildford: British Measurement and Testing Association) Measurement Good Practice Guide No. 36 2003

[24] NASA-HDBK-8739.19-4 2010 *Estimation and Evaluation of Measurement Decision Risk* (Washington, DC: NASA)

[25] NIST 2024 NIST Uncertainty Machine (https://uncertainty.nist.gov/)

[26] NORDTEST 2017 *Handbook for Calculation of Measurement Uncertainty in Environmental Laboratories* 4th edn (Taastrup: NORDTEST) NORDTEST Technical Report 537 Revised

[27] Rishi S 2013 Proposed guidelines for the selection of trapezoidal and triangular distributions for an uncertainty evaluation *NCSL International Measure J. Meas. Sci.* **8** 72–7

IOP Publishing

A Practical Handbook on Measurement Uncertainty
FAQs and fundamentals for metrologists
Swanand Rishi

Appendix A

Coefficients for the Shapiro–Wilk test and the W statistic for various p-values

doi:10.1088/978-0-7503-6462-1ch37

Table A.1. Coefficients. Reproduced from [1], by permission of Oxford University Press.

$n =$	2	3	4	5	6	7	8	9	10	11	12	13	14
a1	0.7071	0.7071	0.6872	0.6646	0.6431	0.6233	0.6052	0.5888	0.5739	0.5601	0.5475	0.5359	0.5251
a2			0.1677	0.2413	0.2806	0.3031	0.3164	0.3244	0.3291	0.3315	0.3325	0.3325	0.3318
a3					0.0875	0.1401	0.1743	0.1976	0.2141	0.2260	0.2347	0.2412	0.2460
a4							0.0561	0.0947	0.1224	0.1429	0.1586	0.1707	0.1802
a5									0.0399	0.0695	0.0922	0.1099	0.1240
a6											0.0303	0.0539	0.0727
a7													0.0240

$n =$	15	16	17	18	19	20	21	22	23	24	25	26
a1	0.5150	0.5056	0.4968	0.4886	0.4808	0.4734	0.4643	0.4590	0.4542	0.4493	0.4450	0.4407
a2	0.3306	0.3290	0.3273	0.3253	0.3232	0.3211	0.3185	0.3156	0.3126	0.3098	0.3069	0.3043
a3	0.2495	0.2521	0.2540	0.2553	0.2561	0.2565	0.2578	0.2571	0.2563	0.2554	0.2543	0.2533
a4	0.1878	0.1939	0.1988	0.2027	0.2059	0.2085	0.2119	0.2131	0.2139	0.2145	0.2148	0.2151
a5	0.1353	0.1447	0.1524	0.1587	0.1641	0.1686	0.1736	0.1764	0.1787	0.1807	0.1822	0.1836
a6	0.0880	0.1005	0.1109	0.1197	0.1271	0.1334	0.1399	0.1443	0.1480	0.1512	0.1539	0.1563
a7	0.0433	0.0593	0.0725	0.0837	0.0932	0.1013	0.1092	0.1150	0.1201	0.1245	0.1283	0.1316
a8		0.0196	0.0359	0.0496	0.0612	0.0711	0.0804	0.0878	0.0941	0.0997	0.1046	0.1089
a9				0.0163	0.0303	0.0422	0.0530	0.0618	0.0696	0.0764	0.0823	0.0876
a10						0.0140	0.0263	0.0368	0.0459	0.0539	0.0610	0.0672
a11								0.0122	0.0228	0.0321	0.0403	0.0476
a12									0.0000	0.0107	0.0200	0.0284
a13											0.0000	0.0094

n =	27	28	29	30	31	32	33	34	35	36	37	38
a1	0.4366	0.4328	0.4291	0.4254	0.4220	0.4188	0.4156	0.4127	0.4096	0.4068	0.4040	0.4015
a2	0.3018	0.2992	0.2968	0.2944	0.2921	0.2898	0.2876	0.2854	0.2834	0.2813	0.2794	0.2774
a3	0.2522	0.2510	0.2499	0.2487	0.2475	0.2463	0.2451	0.2439	0.2427	0.2415	0.2403	0.2391
a4	0.2152	0.2151	0.2150	0.2148	0.2145	0.2141	0.2137	0.2132	0.2127	0.2121	0.2116	0.2110
a5	0.1848	0.1857	0.1864	0.1870	0.1874	0.1878	0.1880	0.1882	0.1883	0.1883	0.1883	0.1881
a6	0.1584	0.1601	0.1616	0.1630	0.1641	0.1651	0.1660	0.1667	0.1673	0.1678	0.1683	0.1686
a7	0.1346	0.1372	0.1395	0.1415	0.1433	0.1449	0.1463	0.1475	0.1487	0.1496	0.1505	0.1513
a8	0.1128	0.1162	0.1192	0.1219	0.1243	0.1265	0.1284	0.1301	0.1317	0.1331	0.1344	0.1356
a9	0.0923	0.0965	0.1002	0.1036	0.1066	0.1093	0.1118	0.1140	0.1160	0.1179	0.1196	0.1211
a10	0.0728	0.0778	0.0822	0.0862	0.0899	0.0931	0.0961	0.0988	0.1013	0.1036	0.1056	0.1075
a11	0.0540	0.0598	0.0650	0.0697	0.0739	0.0777	0.0812	0.0844	0.0873	0.0900	0.0924	0.0947
a12	0.0358	0.0424	0.0483	0.0537	0.0585	0.0629	0.0669	0.0706	0.0739	0.0770	0.0798	0.0824
a13	0.0178	0.0253	0.0320	0.0381	0.0435	0.0485	0.0530	0.0572	0.0610	0.0645	0.0677	0.0706
a14	0.0000	0.0084	0.0159	0.0227	0.0289	0.0344	0.0395	0.0441	0.0484	0.0523	0.0559	0.0592
a15			0.0000	0.0076	0.0144	0.0206	0.0262	0.0314	0.0361	0.0404	0.0444	0.0481
a16					0.0000	0.0068	0.0131	0.0187	0.0239	0.0287	0.0331	0.0372
a17							0.0000	0.0062	0.0119	0.0172	0.0220	0.0264
a18									0.0000	0.0057	0.0110	0.0158
a19											0.0000	0.0053

$n =$	39	40	41	42	43	44	45	46	47	48	49	50
a1	0.3989	0.3964	0.3940	0.3917	0.3894	0.3872	0.3850	0.3830	0.3808	0.3789	0.3770	0.3751
a2	0.2755	0.2737	0.2719	0.2701	0.2684	0.2667	0.2651	0.2635	0.2620	0.2604	0.2589	0.2574
a3	0.2380	0.2368	0.2357	0.2345	0.2334	0.2323	0.2313	0.2302	0.2291	0.2281	0.2271	0.2260
a4	0.2104	0.2098	0.2091	0.2085	0.2078	0.2072	0.2065	0.2058	0.2052	0.2045	0.2038	0.2032
a5	0.1880	0.1878	0.1876	0.1874	0.1871	0.1868	0.1865	0.1862	0.1859	0.1855	0.1851	0.1847
a6	0.1689	0.1691	0.1693	0.1694	0.1695	0.1695	0.1695	0.1695	0.1695	0.1693	0.1692	0.1691
a7	0.1520	0.1526	0.1531	0.1535	0.1539	0.1542	0.1545	0.1548	0.1550	0.1551	0.1553	0.1554
a8	0.1366	0.1376	0.1384	0.1392	0.1398	0.1405	0.1410	0.1415	0.1420	0.1423	0.1427	0.1430
a9	0.1225	0.1237	0.1249	0.1259	0.1269	0.1278	0.1286	0.1293	0.1300	0.1306	0.1312	0.1317
a10	0.1092	0.1108	0.1123	0.1136	0.1149	0.1160	0.1170	0.1180	0.1189	0.1197	0.1205	0.1212
a11	0.0967	0.0986	0.1004	0.1020	0.1035	0.1049	0.1062	0.1073	0.1085	0.1095	0.1105	0.1113
a12	0.0848	0.0870	0.0891	0.0909	0.0927	0.0943	0.0959	0.0972	0.0986	0.0998	0.1010	0.1020
a13	0.0733	0.0759	0.0782	0.0804	0.0824	0.0842	0.0860	0.0876	0.0892	0.0906	0.9190	0.0932
a14	0.0622	0.0651	0.0677	0.0701	0.0724	0.0745	0.0765	0.0783	0.0801	0.0817	0.0832	0.0846
a15	0.0515	0.0546	0.0575	0.0602	0.0628	0.0651	0.0673	0.0694	0.0713	0.0731	0.0748	0.0764
a16	0.0409	0.0444	0.0476	0.0506	0.0534	0.0560	0.0584	0.0607	0.0628	0.0648	0.0667	0.0685
a17	0.0305	0.0343	0.0379	0.0411	0.0442	0.0471	0.0497	0.0522	0.0546	0.0568	0.0588	0.0608
a18	0.0203	0.0244	0.0283	0.0318	0.0352	0.0383	0.0412	0.0439	0.0465	0.0489	0.0511	0.0532
a19	0.0101	0.0146	0.0188	0.0227	0.0263	0.0296	0.0328	0.0357	0.0385	0.0411	0.0436	0.0459
a20	0.0000	0.0049	0.0094	0.0136	0.0175	0.0211	0.0245	0.0277	0.0307	0.0335	0.0361	0.0386
a21			0.0000	0.0045	0.0087	0.0126	0.0163	0.0197	0.0229	0.0259	0.0288	0.0314
a22					0.0000	0.0042	0.0081	0.0118	0.0153	0.0185	0.0215	0.0244
a23							0.0000	0.0039	0.0076	0.0111	0.0143	0.0174
a24									0.0000	0.0037	0.0071	0.0104
a25											0.0000	0.0035

Table A.2. W statistics. Reproduded from [1], by permission of Oxford University Press

$n \backslash p$	0.01	0.02	0.05	0.1	0.5	0.9	0.95	0.98	0.99
3	0.753	0.756	0.767	0.789	0.959	0.998	0.999	1.000	1.000
4	0.687	0.707	0.748	0.792	0.935	0.987	0.992	0.996	0.997
5	0.686	0.715	0.762	0.806	0.927	0.979	0.986	0.991	0.993
6	0.713	0.743	0.788	0.826	0.927	0.974	0.981	0.986	0.989
7	0.730	0.760	0.803	0.838	0.928	0.972	0.979	0.985	0.988
8	0.749	0.778	0.818	0.851	0.932	0.972	0.978	0.984	0.987
9	0.764	0.791	0.829	0.859	0.935	0.972	0.978	0.984	0.986
10	0.781	0.806	0.842	0.869	0.938	0.972	0.978	0.983	0.986
11	0.792	0.817	0.850	0.876	0.940	0.973	0.979	0.984	0.986
12	0.805	0.828	0.859	0.883	0.943	0.973	0.979	0.984	0.986
13	0.814	0.837	0.866	0.889	0.945	0.974	0.979	0.984	0.986
14	0.825	0.846	0.874	0.895	0.947	0.975	0.980	0.984	0.986
15	0.835	0.855	0.881	0.901	0.950	0.975	0.980	0.984	0.987
16	0.844	0.863	0.887	0.906	0.952	0.976	0.981	0.985	0.987
17	0.851	0.869	0.892	0.910	0.954	0.977	0.981	0.985	0.987
18	0.858	0.874	0.897	0.914	0.956	0.978	0.982	0.986	0.988
19	0.863	0.879	0.901	0.917	0.957	0.978	0.982	0.986	0.988
20	0.868	0.884	0.905	0.920	0.959	0.979	0.983	0.986	0.988
21	0.873	0.888	0.908	0.923	0.960	0.980	0.983	0.987	0.989
22	0.878	0.892	0.911	0.926	0.961	0.980	0.984	0.987	0.989
23	0.881	0.895	0.914	0.928	0.962	0.981	0.984	0.987	0.989
24	0.884	0.898	0.916	0.930	0.963	0.981	0.984	0.987	0.989
25	0.888	0.901	0.918	0.931	0.964	0.981	0.985	0.988	0.989
26	0.891	0.904	0.920	0.933	0.965	0.982	0.985	0.988	0.989
27	0.894	0.906	0.923	0.935	0.965	0.982	0.985	0.988	0.990
28	0.896	0.908	0.924	0.936	0.966	0.982	0.985	0.988	0.990
29	0.898	0.910	0.926	0.937	0.966	0.982	0.985	0.988	0.990
30	0.900	0.912	0.927	0.939	0.967	0.983	0.985	0.988	0.990
31	0.902	0.914	0.929	0.940	0.967	0.983	0.986	0.988	0.990
32	0.904	0.915	0.930	0.941	0.968	0.983	0.986	0.988	0.990
33	0.906	0.917	0.931	0.942	0.968	0.983	0.986	0.989	0.990
34	0.908	0.919	0.933	0.943	0.969	0.983	0.986	0.989	0.990
35	0.910	0.920	0.934	0.944	0.969	0.984	0.986	0.989	0.990
36	0.912	0.922	0.935	0.945	0.970	0.984	0.986	0.989	0.990
37	0.914	0.924	0.936	0.946	0.970	0.984	0.987	0.989	0.990
38	0.916	0.925	0.938	0.947	0.971	0.984	0.987	0.989	0.990
39	0.917	0.927	0.939	0.948	0.971	0.984	0.987	0.989	0.991
40	0.919	0.928	0.940	0.949	0.972	0.985	0.987	0.989	0.991
41	0.920	0.929	0.941	0.950	0.972	0.985	0.987	0.989	0.991
42	0.922	0.930	0.942	0.951	0.972	0.985	0.987	0.989	0.991
43	0.923	0.932	0.943	0.951	0.973	0.985	0.987	0.990	0.991
44	0.924	0.933	0.944	0.952	0.973	0.985	0.987	0.990	0.991

(*Continued*)

Table A.2. (*Continued*)

$n \backslash p$	0.01	0.02	0.05	0.1	0.5	0.9	0.95	0.98	0.99
45	0.926	0.934	0.945	0.953	0.973	0.985	0.988	0.990	0.991
46	0.927	0.935	0.945	0.953	0.974	0.985	0.988	0.990	0.991
47	0.928	0.936	0.946	0.954	0.974	0.985	0.988	0.990	0.991
48	0.929	0.937	0.947	0.954	0.974	0.985	0.988	0.990	0.991
49	0.929	0.939	0.947	0.955	0.974	0.985	0.988	0.990	0.991
50	0.930	0.938	0.947	0.955	0.974	0.985	0.988	0.990	0.991

Reference

[1] Shapiro S S and Wilk M B 1965 An analysis of variance for normality (complete samples) *Biometrika* **52** 591–611

A Practical Handbook on Measurement Uncertainty
FAQs and fundamentals for metrologists
Swanand Rishi

Appendix B

Some useful web sites

1. Agilent Technologies: www.agilent.com
2. American National Standards Institute (ANSI): www.ansi.org
3. American Society for Quality (ASQ): https://asq.org
4. American Society for Testing and Materials (ASTM): www.astm.org
5. American Society of Mechanical Engineers (ASME): www.asme.org
6. Bureau International des Poids et Mesures (BIPM): www.bipm.org
7. Co-Operation on International Traceability in Analytical Chemistry (CITAC): www.citac.ws
8. EURACHEM: www.eurachem.org
9. European Co-operation for Accreditation (EA): www.european-accreditation.org
10. European Federation of National Associations of Measurement, Testing and Analytical Laboratories (EUROLAB): www.eurolab.org.
11. Fluke Corporation: www.fluke.com
12. International Electrotechnical Commission (IEC): www.iec.ch
13. International Laboratory Accreditation Cooperation (ILAC): www.ilac.org
14. International Organization for Standardization (ISO): www.iso.ch
15. International Organization of Legal Metrology (OIML): www.oiml.org
16. International Union of Pure and Applied Chemistry (IUPAC): www.iupac.org
17. International Union of Pure and Applied Physics (IUPAP): www.iupap.org
18. National Accreditation Board for Testing and Calibration Laboratories (NABL): www.nabl-india.org
19. National Association of Testing Authorities, Australia (NATA): www.nata.com.au
20. National Conference of Standards Laboratories International: www.ncslin-ternational.org
21. National Institute of Standards and Technology (NIST): www.nist.gov
22. National Physical Laboratory, India: www.nplindia.org

23. National Physical Laboratory, UK: www.npl.co.uk
24. NIST Uncertainty Machine: uncertainty.nist.gov
25. NORDTEST (Association of Nordic Laboratories): www.nordtest.org
26. Physikalisch-Technische Bundesanstalt (PTB, Germany): www.ptb.de
27. United Kingdom Accreditation Service (UKAS): www.ukas.com

www.ingramcontent.com/pod-product-compliance
Lightning Source LLC
Chambersburg PA
CBHW080545220326
41599CB00032B/6369